The Entangled Brain

How Perception, Cognition, and Emotion

Luiz Pessoa

The MIT Press
Cambridge, Massachusetts
London, England

The MIT Press would like to thank the anonymous peer reviewers who provided comments on drafts of this book. The generous work of academic experts is essential for establishing the authority and quality of our publications. We acknowledge with gratitude the contributions of these otherwise uncredited readers.

This book was set in Stone Serif and Stone Sans by Westchester Publishing Services. Printed and bound in the United States of America.

Library of Congress Cataloging-in-Publication Data

Names: Pessoa, Luiz, author.
Title: The entangled brain : how perception, cognition, and emotion are woven together / Luiz Pessoa.
Description: Cambridge, Massachusetts : The MIT Press, [2022] | Includes bibliographical references and index.
Identifiers: LCCN 2021061878 (print) | LCCN 2021061879 (ebook) | ISBN 9780262544603 (paperback) | ISBN 9780262372107 (pdf) | ISBN 9780262372114 (epub)
Subjects: LCSH: Perception. | Emotions and cognition. | Brain. | Neuropsychology.
Classification: LCC BF311 .P3767 2022 (print) | LCC BF311 (ebook) | DDC 153—dc23/eng/20220411
LC record available at https://lccn.loc.gov/2021061878
LC ebook record available at https://lccn.loc.gov/2021061879

10 9 8 7 6 5 4

The Entangled Brain

To Amelie and Meg

Contents

Preface

Good books by active scientists offer serious treatments of sophisticated topics in biology, evolution, physics, and mathematics. Since I was a teenager, such books have been a major inspiration to me. In *The Entangled Brain*, I wanted to do the same—hopefully!—and write about my favorite topic, which has been the object of my research during the past three decades. Many good, and even excellent, general-audience books about neuroscience focus on a certain aspect of the mind-brain—say, addiction, cognition, memory, and so on. This makes sense, as there is probably too much to cover and the topic is better suited for a college textbook, for example. But I wanted to write about the brain broadly construed, not a narrower subject such as "emotion" or "reward."

That's what I tried to do here. But this is a relatively short book, not a 1,000-page tome. So, it had to be selective and leave a lot out. That also meant that it had to be more idiosyncratic, reflecting my view of the brain and not necessarily providing a more extensive exposition with pros and cons of many concepts and ideas. This decision entailed following what I consider a way of thinking about the brain that, while not rare, is also *not* mainstream among neuroscientists.

A central thesis of the book is that biology does not work like physics, and even less so like engineering. Biological systems are not easily reducible to separate units that, when put together, give us the whole back. Unfortunately, in my view, even brain scientists (many of them, at least) don't fully appreciate this idea. Their descriptions of the brain are full of labels for brain regions, indicating that they perform function X (here's "fear") or Y (here's "reward"), as if the separate pieces functioned quasi-autonomously. This thinking reflects a mapping between structure (anatomy) and function (behavior) that goes back more than a century. For example, in the early twentieth century, Korbinian Brodmann, an early neuroanatomist, subdivided the human brain into roughly 50 specific anatomical units that were thought to map to relatively distinct functions. To this day, Brodmann's map and its refinements are routinely used by researchers. Indeed, one of the central approaches of neuroscience has been a divide-and-conquer strategy that tries to break up the entire organ into subcomponents that can be, purportedly, properly understood. They then can be put back together in the hope that the overall functionality will reflect the summed individual parts. I believe this strategy is problematic; in fact, it is inadequate to understand systems like the brain (and genetics by the way) in which the interactions among the parts create mechanisms and processes than cannot be derived by looking at parts in isolation.

In *The Entangled Brain*, I wanted to avoid what I find in many general-audience books—namely, descriptions that simplify the brain to such an extreme as to appear, at times, caricatures. For example, in the context of emotion and motivation, an often-heard narrative is that primitive, subcortical regions like the amygdala (presumably "responsible for fear") and the striatum (presumably "responsible for reward") produce automatic behaviors that are next-to-impossible to subvert—hence, anxiety disorders and addiction. At the same time, the prefrontal cortex, the "newer and more rational" part of the brain, allows us to exert control over the subcortical bits and correct behaviors when appropriate (no cake-eating if one is on a diet, as an example). The treatment in the pages that follow adheres to a way of thinking that eschews these first-order explanations. The resulting story is not so simple, but I believe readers are more than ready to face the complexity. We don't have to put functions inside little boxes in the brain and tell neat stories. Reality is immensely more complex.

The view formulated here is that parts of the brain work in a coordinated fashion, such that functions are carried out by large-scale distributed circuits, also called large-scale *networks*. In other words, collections of gray matter parts exchange signals with one another and, by doing so, bring about behaviors. The circuits are distributed, not local, involving disparate parts in the cortex and the subcortex, for example. And they are "large scale" because they don't only involve a pair, or possibly a few regions, but many components working simultaneously. That is the sense in which *the brain is entangled*, as summarized by the book's title. The overall goal of the book is to introduce the central nervous system to readers in a sophisticated yet engaging manner—I hope! The text exposes readers to some of the complexities surrounding our understanding of the brain, without submitting the reader to a tsunami of technicalities.

Many lay readers (and some neuroscientists) implicitly assume that the human brain is rather unique with its prominent cortex (a word that means "bark" in Latin, like the exterior covering of the trunk and branches of a tree). However, in the past several decades, neuroanatomists have uncovered striking similarities in the overall "plan" of the brain of all vertebrates (fish, amphibians, reptiles, birds, and mammals). It is therefore both helpful and important to understand the central nervous system of humans from this so-called *comparative* lens. I cover this evolutionary link in chapter 9.

The concept of *complex systems* permeates *The Entangled Brain*. Complex systems are comprised of many relatively simple interacting parts and exhibit *emergent* behaviors: properties absent at the level of the individual parts but observed for the system as a whole. Starting in the 1940s with intellectual movements such as cybernetics and systems biology, complex systems theory has spread into most fields of knowledge where the interactions between elements (including feedback loops) challenge our ability to decipher how a given system works. Today, many fields including neuroscience, ecology, and the study of evolution apply insights from this domain of knowledge.

The pages ahead introduce the reader to the brain at the systems level. The book does not aim to be comprehensive. Neuroscience is such a sprawling field of research that this is not really possible. Whereas the text does not spell out a "novel" view of the brain per se, it closely reflects a line of thinking that I have outlined, and continue to develop, in a series of peer-reviewed conceptual papers, as well as published papers on "cognitive-emotional interactions" from a comparative viewpoint (that is, when all vertebrates are considered). In particular, one notion that is the outcome of my research over the past 30 years—and that undergirds the entire text—is that perception, cognition, action, emotion, and cognition are closely interrelated in the brain. You can't point to the brain and say, "This is where X happens."

Many books are the product of long years of work, on and off. Early drafts of this book date to the end of 2016. The text took much of its present form during a couple delightful months in Turin, Italy, where on a short sabbatical stay I taught a course based on the material. I am grateful to the University of Turin for hosting me, in particular my friend Marco Tamietto for his amazing hospitality. I am also grateful to the University of Maryland, College Park, for the sabbatical support. I'm thankful to the students at the University of Turin for the feedback on my lectures and draft chapters—*grazie mille*. Marco Viola also provided excellent feedback on chapter 4 (What Do Brain Areas Do?). I'm also thankful for the feedback that I received from the students in my class at the University of Maryland, when I taught a follow-up version of the original course at Turin. I am indebted to Loreta Medina and Ester Desfilis for sharing their enormous knowledge of vertebrate neuroanatomy over the past years; their guidance has helped shape my (evolving!) view of the organization of the brain. I wish to thank Sydni Roberts, Kelly Morrow, Govinda Surampudi, and especially Trang Nguyen for their help with the figures in this book; Trang also helped a lot with the index. Feedback of colleagues on Twitter was also key in finalizing the book's title.

I'm very grateful to my colleague Michael Anderson for putting me in touch with Phil Laughlin at MIT Press, who was enthusiastic about the project right away. The peer reviewers for MIT Press also made several helpful suggestions; thank you.

Finally, I thank Meg and Amelie for all their love.

1 From One Area at a Time to Networked Systems

We begin our journey into how the brain brings about the mind: our perceptions, actions, thoughts, and feelings. Historically, the study of the brain has proceeded in a divide-and-conquer way, trying to figure out the function of individual areas—chunks of gray matter that contain neurons—one at a time. This book makes the case that because the brain is not a modular system, we need conceptual tools that can help us decipher how highly *entangled*, complex systems function.

In 2016, a group of investigators published a map of the major subdivisions of the human cerebral cortex—the outer part of the brain—in the prestigious journal *Nature* (figure 1.1). The partition delineated 180 areas in each hemisphere (360 in total), each of which represents a unit of "architecture, function, and connectivity" (Glasser et al., 2016, 171).[1] Many researchers celebrated the new result given the long-overdue need to replace the de facto standard called the "Brodmann map." Published in 1908 by Korbinian Brodmann, the map describes approximately 50 areas in each hemisphere (100 in total) based on local features, such as cell type and density, that Brodmann discovered under the microscope.

Notwithstanding the need to move past a standard created before the First World War, the 2016 cartographic description builds on an idea that was central to earlier efforts: Brain tissue should be understood in terms of a set of well-defined, spatially delimited sectors. Thus the concept of a brain *area* or *region*:[2] a unit that is both anatomically and functionally meaningful. The notion of an area/region is at the core of neuroscience as a discipline, with its central challenge of unraveling how behaviors originate from cellular matter. Put another way, how does function (manifested externally by behaviors) relate to structure (such as different neuron types and their arrangement)? How do groups of neurons—the defining cell type of the brain—lead to sensations and actions?

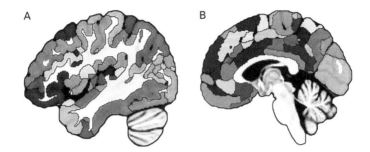

Figure 1.1
Map of brain areas of the cortex published in 2016. Each hemisphere (or half of the brain) contains 180 areas indicated by different shades of gray and outlines. (a) The brain is shown in a side view. (b) The brain is shown through a cut revealing the middle. *Source*: Regions as defined by Glasser et al. (2016).

As a large and heterogeneous collection of neurons and other cell types, the central nervous system—including cortical and subcortical parts—is a formidably complex organ. (The cortex is the outer surface with grooves and bulges; the subcortex comprises other cell masses that sit underneath. We'll go over the basics of brain anatomy in chapter 2.) To unravel how it works, some strategy of *divide and conquer* seems to be necessary. How else can the brain be understood without breaking it down into subcomponents? But this approach also exposes a seemingly insurmountable chicken-and-egg problem: If we don't know how it works, how can we determine the "right" way to subdivide it? Finding the proper unit of function, then, has been at the center of the quest to crack the mind-brain problem.

Historically, two winners in the search for rightful units have been the neuron and the individual brain area. At the cellular level, the neuron reigns supreme. Since the work of Ramon y Cajal,[3] the Spanish scientific giant who helped establish neuroscience as an independent discipline, the main cell type of the brain is considered to be the neuron (and neurons come in many varieties, both in terms of morphology and physiology). These cells communicate with one another through electrochemical signaling. If they are sufficiently excited by other neurons, their equilibrium voltage changes and they generate a "spike": an electrical signal that propagates along the neuron's thin extensions (called axons), much like a current flowing through a wire. The spike from a neuron can then influence downstream neurons. And so on.

At the supra-cellular level, the chief unit is the *area*. But what constitutes an area? Dissection techniques and the study of neuroanatomy during the European Renaissance were propelled to another level by Thomas Willis's monumental *Cerebri anatome* published in 1664. The book rendered in exquisite detail the morphology of the human brain, including detailed drawings of subcortical structures and the cerebral hemispheres containing the cortex. For example, Willis described a major structure of the subcortex, the striatum, that I will discuss at length in the chapters to follow. With time, as anatomical methods improved with more powerful microscopes and diverse stains (which mark the presence of chemical compounds in the cellular milieu), more and more subcortical areas were discovered. In 1819, the German anatomist Karl Burdach described a mass of gray matter that could be seen in slices through the temporal lobe. He called the structure the "amygdala"[4]—given that it is shaped like an almond ("amygdala" means almond in Latin)—now famous for its contributions to fear processes. And techniques developed in the second half of the twentieth century revealed that it is possible to delineate a least a dozen subregions within its overall territory.

The seemingly benign question—what counts as an area?—is far from straightforward. For instance, is the amygdala one region or 12 regions? This region is far from an esoteric case. All subcortical areas have multiple subdivisions, and some have boundaries that are more like fuzzy zones than clearly defined lines. The challenges of partitioning the cortex, the outer laminated mantle of the cerebrum, are enormous, too. That's where the work of Brodmann and others, and more recently the research that led to the 180-area parcellation (figure 1.1), comes in. These developments introduce a set of criteria to subdivide the cortex into constituent parts. For example, although neurons in the cortex are arranged in a layered fashion, the number of cell layers can vary. Therefore, identifying a transition between two cortical sectors is aided by differences in cell density and layering.

How Modular Is the Brain? Not Much at All

When subdividing a larger system—one composed of lots of parts—the concept of *modularity* comes into play. Broadly speaking, modularity refers to the degree of interdependence of the many parts that comprise the system of interest. On the one hand, a *decomposable* system is one in which each subsystem operates according to its own intrinsic principles, independently of the others—we say that this system is highly modular. On the other hand, a *non*-decomposable system is one in which the connectivity and interrelatedness of the parts is such that they are no longer clearly separable. Whereas the two extremes serve as useful anchors to orient our thinking, in practice one finds a continuum of possible organizations, so it's more useful to think of the degree of modularity of a system.

Science as a discipline is inextricably associated with understanding entities in terms of a set of constituent subparts. Neuroscience has struggled with this modus operandi since its early days, and debates about "localizationism" versus "connectionism"—how local or how interconnected brain mechanisms are—have always been at the core of the discipline. By and large, a fairly *modular* view has prevailed in neuroscience. Fueled by a reductionistic drive that has served science well, most investigators have formulated the study of the brain as a problem of dissecting the multitude of "sub-organs" that make it up. To be true, brain parts are not viewed as isolated islands and are understood to communicate with one another. But, commonly, the plan of attack assumes that the nervous system is decomposable[5] in a meaningful way in terms of patches of tissue (as in figure 1.1) that perform well-defined computations—if only we can determine what they are.

There have been proposals of non-modular processing, too. The most famous example is that of Karl Lashley who, starting in the 1930s, defended the idea of "cortical equipotentiality"—namely, that most of the cortex functions jointly, as a unit. Thus, the extent of a behavioral deficit caused by a lesion depended on the amount of cortex that was compromised—small lesions cause small deficits, large lesions cause larger ones. Although Lashley's proposal was clearly too extreme and rejected empirically, historically, many ideas of decentralized processing have been entertained by neuroscientists. Let's discuss some of their origins.

The Entangled Brain

The field of *artificial intelligence* (AI) is said to have been born at a workshop at Dartmouth College in 1956. Early AI focused on the development of computer algorithms that could emulate human-like "intelligence," including simple forms of problem solving, planning, knowledge representation, and language understanding. A parallel and competing approach—what was to become the field of *artificial neural networks*, or neural networks, for short—took its inspiration instead from *natural* intelligence and adopted basic principles of the biology of nervous systems. In this non-algorithmic framework, collections of simple processing elements work together to execute a task. An early example was the problem of pattern recognition, such as recognizing sequences of 0s and 1s. A more intuitive, modern application addresses the goal of image classification. Given a set of pictures coded as a collection of pixel intensities, the task is to generate an output that signals a property of interest; say, output "1" if the picture contains a face and "0" otherwise. The underlying idea behind artificial neural networks was that "intelligent" behaviors result from the joint operation of simple processing elements, like artificial neurons that sum their inputs and generate an output if the sum exceeds a certain threshold value. We'll discuss neural networks again in chapter 8, but here we emphasize their conceptual orientation: thinking of a system in terms of *collective computation*.

The 1940s and 1950s were also a time when, perhaps for the first time, scientists started systematically developing *theories of systems* generally conceived. The intellectual *cybernetics* movement was centrally concerned with how systems regulate themselves so as to remain within stable regimes; for example, normal, awake human temperature remains within a narrow range, varying less than a degree Celsius. *Systems theory*, also called general systems theory or complex systems theory, tried to formalize how certain properties might originate from the interactions of multiple, and possibly simple, constituent parts. How does "wholeness" come about in a way that is not immediately explained by the properties of the parts?

Fast-forward to 1998 and the publication of a paper entitled "Collective Dynamics of 'Small-World' Networks" (Watts and Strogatz 1998). The study proposed that the organization of many biological, technological, and social networks gives them enhanced signal-propagation speed, computational power, and synchronization among parts. And these properties are possible even in systems where most elements are connected locally, with only some elements having "arbitrary" connections. (For example, consider a network of interlinked computers, such as the internet. Most computers are only connected to others in a fairly local manner—say, within a given department within a company or university. However, a few computers have connections to other computers that are geographically quite far.)

Duncan Watts and Steven Strogatz applied their techniques to study the organization of a social network containing more than 200,000 actors. As we'll discuss in chapter 10, to make a "network" out of the information they had available, they considered two actors to be "connected" if they had appeared in a film together. Although a given actor was only connected to a small number of other performers (around 60), they discovered that it was possible to find short "paths" between any two actors. (The path A–B–C links actors A and C, which have not participated in the same film, if both of them have co-acted with actor B.) Remarkably, on average, paths containing only four connections (such as the path A–B–C–D–E linking actors A and E) separated a given pair of actors picked at random from the set of 200,000. The investigators dubbed this property "small world" by analogy with the popularly known idea of "six degrees of separation" and suggested that it is a hallmark of many types of networks—one can travel from A to Z very expediently.

The paper by Watts and Strogatz, and a related paper by Albert-László Barabási and Réka Albert that appeared the following year (1999), set off an avalanche of studies on what has become known as "network science"— the study of interconnected systems comprised of more elementary components, such as a social network of individual persons. This research field has grown enormously since then, and novel techniques are actively being applied to social, biological, and technological problems to refine our view of "collective behaviors." These ideas resonated with research in brain science, too, and it didn't take long before investigators started applying network science techniques to study their data. This was particularly the case in human neuroimaging, which employs magnetic resonance imaging (MRI) scanners to measure activity throughout the brain during varied experimental conditions. Network science provides a spectrum of analysis tools to tackle brain data. First and foremost, the framework encourages researchers to conceptualize the nervous system in terms of network-level properties. That is to say, whereas individual parts—brain areas or other such units— are important, collective or system-wide properties must be targeted.

Neuroscientific Explanations

Neuroscience seeks to answer the following central question: How does the brain generate behavior?[6] Broadly speaking, there are three types of study: lesion, activity, and manipulation. *Lesion* studies capitalize on naturally occurring injuries, including those caused by tumors and vascular accidents; in nonhuman animals, precise lesions can be created surgically, thus allowing much better control over the affected territories. What types of behavior are affected by such lesions? Perhaps patients can't pay attention to visual information the way they used to, or maybe they have difficulty moving a limb. *Activity* studies measure brain signals. The classic technique is to insert a microelectrode into the tissue of interest and measure electrical signals in the vicinity of neurons (it is also possible to measure signals inside a neuron itself, but such experiments are more technically challenging). Voltage changes provide an indication of a change in state of the neuron(s) closest to the electrode tip. And determining how such changes are tied to the behaviors performed by an animal provides clues about how they contribute to them. *Manipulation* studies directly alter the state of the brain by either silencing or enhancing signals. Again, the goal is to see how sensations and actions are affected.

Although neuroscience studies are incredibly diverse, one way to summarize them is as follows: "Area or circuit X is *involved* in behavior Y" (where a circuit is a group of areas). A lesion study might determine that patients with damage to the so-called cortex of the anterior insula have the ability to quit smoking easily, without relapse, leading to the conclusion that the insula is a critical substrate in the addiction to smoking (Naqvi et al. 2007). Why? Quitting is hard in general, of course. But it turns out to be easy if one's anterior insula is nonfunctional. It is logical, therefore, to surmise that, when intact, this region's operation somehow promotes addiction. An activation study using functional MRI might observe stronger signals in parts of the visual cortex when participants view pictures of faces compared to when they are shown many kinds of pictures that don't contain faces (pictures of multiple types of chairs, shoes, or other objects.). This could suggest that this part of the visual cortex is important for the perception of faces. A manipulation study that enhances activity in the prefrontal cortex in monkeys might observe an improvement in tasks that requires careful attention to visual information.

Many journals require "significance statements" in which authors summarize the importance of their studies to a broader audience. In the instances of the previous paragraph, the authors could say something like this: (1) The insula *contributes* to conscious drug urges and to decision-making processes that precipitate relapse; (2) the fusiform gyrus (the particular area of visual cortex that responds vigorously to faces) is *involved* in face perception; and (3) the prefrontal cortex *enhances* performance of behaviors that are challenging and require attention.

The examples above weren't gratuitous; all were important studies published in very respected scientific journals. Although these were rigorous experimental studies, they don't quite inform about the underlying mechanisms.[7] In fact, if one combs the peer-reviewed literature, one finds a plethora of *filler terms*[8]—words like "contributes," "involved," and "enhances" above (figure 1.2)—that stand in for the processes we presume did the "real" work. This is because, by and large, neuroscience studies don't sufficiently determine, or even strongly constrain, the underlying mechanisms that link brain to behavior.

Scientists strive to discover the mechanisms supporting the phenomena they study. But what precisely is a *mechanism*? Borrowing from the philosopher William Bechtel, it can be defined as "a structure performing a function in virtue of its parts, operations, and/or organization. The functioning of the mechanism is responsible for one or more phenomena" (Bechtel 2008, 13). Rather abstract, of course, but in essence it means *how something happens*. The more clear-cut we can be about it, the better. For example, in physics, precision actually involves mathematical equations. Note that mechanisms and explanations are always at some *level of explanation*. A typical explanation about combustion motors in automobiles will invoke pistons, fuel, or controlled explosions. It will not discuss these phenomena in term of particle physics, for instance; it won't invoke electrons, protons, or neutrons.

Filler verbs used in neuroscience explanations

Reflects	Encodes	Reveals	Induces
Involves	Enables	Regulates	Ensures
Mediates	Supports	Generates	Promotes
Modulates	Determines	Shapes	Plays a role in
Contributes to	Underlies	Produces	Is associated with

Figure 1.2
Because little is known about how brain mechanisms bring about behaviors, neuroscientists use "filler" verbs, most of which add relatively little substantive content to the statements made.
Source: List of words from Krakauer et al. (2017).

We currently *lack an understanding* of most brain science phenomena. Therefore, when an experiment finds that changes occur in, say, the amygdala during classical aversive conditioning (learning that a once-innocuous stimulus is now predictive of a shock; see chapter 5), we might find that cell responses there increase in parallel to the observed behavior—as the behavior is acquired, cells responses concomitantly increase. Although this is a very important finding, it remains relatively shallow in clarifying what's going on. Of course, if through a series of studies we come to discern how amygdala activity increases, decreases, or stays the same when learning changes accordingly, we are closer to legitimately saying that we grasp the underlying mechanisms.

Pleading Ignorance

How much do we know about the brain today? In the media, there is no shortage of news about novel discoveries explaining why we are all stressed, overeat, or cannot stick to our resolutions for the new year. General-audience books on brain and behavior are extremely popular, even if we don't count the ever-ubiquitous self-help books, which are themselves loaded with purported insights from brain science. And judging from the size of graduate school textbooks (some of which are even hard to lift), current knowledge is a deep well.

In reality, we know rather little. What we've learned barely scratches the surface.

Consider, for example, this statement by eminent neuroscientists: "Despite centuries of study of brain–behavior relationships, a clear formalization of the function of many brain regions, accounting for the engagement of the region in different behavioral functions, is lacking" (Genon et al. 2018, 362).[9] A clear-headed description of our state of ignorance was given by Ralph Adolphs and David Anderson, both renowned professors at the California Institute of Technology, in their book *The Neuroscience of Emotion*:

> We can predict whether a car is moving or not, and how fast it is moving, by "imaging" its speedometer. That does not mean that we understand how an automobile works. It just means that we've found something that we can measure that is strongly correlated with an aspect of its function. Just as with the speedometer, imaging [measuring] activity in the amygdala (or anywhere else in the brain), in the absence of further knowledge, tells us nothing about the causal mechanism and only provides a "marker" that may be correlated with an emotion. (Adolphs and Anderson 2018, 31)

Although these authors were discussing the state of knowledge regarding emotion and the brain, it is fair to say that their summary applies to neuroscience more generally—the science of brain and behavior is still in its (very) early days.

The gap—no, gulf—between scientific knowledge and how it is portrayed by the general media is sizable indeed. We may encounter not only a piece in a popular magazine found in a medical office, but a serious article in, say, the *New York Times* or *The Guardian* newspapers of some heft. The problem even extends to most science communication books, especially those with a more clinical or medical slant.

Mechanisms and Complexity in Biology

How does something work? As discussed above, science approaches this question by trying to work out *mechanisms*. We seek "machine-like" explanations, much like describing how an old, intricate clock functions. Consider a Rube Goldberg apparatus (for an example, see figure 1.3), accompanied by directions on how to use it to light a bulb:[10]

- The boxing glove is triggered.
- The glove punches the bowling ball, which slides down and knocks the pin.
- The bowling pin pulls a string that opens the birdcage door (releasing the bird!) and tilts the wood plank, which makes the billiard ball go down the ramps.
- The billiard ball hits the closest brick, triggering a domino effect that knocks down all bricks.

  . . .

- The hammer hits the hand, which falls and, in so doing, pulls the cord.
- The light bulb lights up!

The "explanation" above works because it provides a *causal* narrative: a series of cause-and-effect steps that slowly but surely lead to the outcome. Although this example is artificial of course (no one would light a bulb like that), it epitomizes a style of explanation that is the gold standard of science.

Yet, biological phenomena frequently involve complex, tangled webs of explanatory factors. Consider guppies, small fish native to streams in South America, which show geographical variation in many kinds of traits,

Figure 1.3
Rube Goldberg apparatus as an example of mechanical explanation.

including color patterns.[11] To explain the morphological and behavioral variation among guppies, the biologist John Endler suggested that we consider a "network of interactions" (figure 1.4). The key point was not to focus on the details of the interactions, but the fact that they exist. Complex as it may look, Endler's network is "simple" as far as biological systems go. It doesn't involve bidirectional influences (double-headed arrows), that is, those in which A affects B and B affects A in turn (see chapter 8). Still, most biological systems are organized like that.

Contrast such state of affairs to the vision encapsulated by Isaac Newton's statement that "truth is ever to be found in simplicity, and not in the multiplicity and confusion of things" (Mazzocchi 2008, 10).[12] This stance is such an integral part of the canon of science as to constitute a form of First Commandment. Newton himself was building on the shoulders of René Descartes, the French polymath who helped canonize *reductionism* (see chapter 4) as part of Western thinking and philosophy. To him, the world was to be regarded as a clockwork mechanism. That is to say, in order to understand something, it is necessary to investigate the parts and

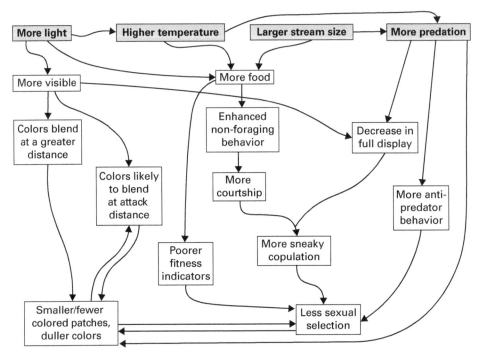

Figure 1.4
Multiple explanatory factors that influence morphological and behavioral variation among South American guppies, illustrating the rich network of relationships. Two of the relationships are bidirectional (see parallel arrows).
Source: Figure inspired by Endler (1995).

then reassemble the components to again create the whole—the essence of reductionism. Fast-forward to the second half of the twentieth century. The dream of extending the successes of the Cartesian mindset captivated biologists. As Francis Crick, one of the co-discoverers of the structure of DNA put it, "The ultimate aim of the modern movement in biology is to explain all biology in terms of physics and chemistry" (Mazzocchi 2008, 10). Reductionism indeed.

So, where is neuroscience today? The mechanistic tradition set off by Newton's *Principia Mathematica*—arguably the most influential scientific achievement ever—is a major conceptual driving force behind how brain scientists think. Although many biologists view their subject matter as different from physics, for example, scientific practice is very much dominated by a mechanistic approach. The present book embraces a different way of thinking, one that revolves around ideas of "collective phenomena," ideas of networks, and ideas about complexity. The book is as much about what we know about the brain as it is a text to stimulate how we can think about the brain as a highly complex network—indeed entangled—system.

Before we can start our journey, we need to define a few terms and learn a little about anatomy.

Road Ahead

If the central nervous system is indeed highly networked, then after learning some anatomy, you might expect that a natural subsequent chapter would be one on *complex systems*, including some introduction to *network science*. This was *not* the approach I took, however. Instead, my aim was to *build up the need* to consider the brain as a complex, entangled system. Thus, chapters 2 to 7 consider brain regions in a fair amount of isolation. In part, this is done so the reader can, in fact, gradually appreciate the limitations of this way of thinking. The approach is also pragmatic because the field of neuroscience has tended to study parts of the brain separately. So, we will use the knowledge accrued as a starting point, without forgetting the need to move beyond this type of description.

Before we get to chapter 8, which discusses complex systems, we have a few stops along the way. We start with some neuroanatomy (chapter 2), as we must learn how to orient ourselves around the target territory and learn its overall organization. In chapter 3, I describe the idea of a hypothetical "minimal brain" that allows an animal to defend itself and seek rewards, essential components of survival. How do sensations lead to actions via simple sensorimotor circuits? We'll see that action flexibility necessitates uncoupling sensory and motor components. The objectives of chapter 3 are twofold. First, it is meant to introduce you to some brain regions/sectors and some of the functions they contribute to. (The linguistic construction of the previous sentence—"functions they contribute to"—may seem a poor one, but it is a consequence of the following central notion: Regions *contribute to* functions; by and large, they don't have isolated functions.) Second, the intent is to show that the business of a brain region, according to the view espoused here, needs to be situated in the context of multi-region circuits: What does a brain region do *in combination* with other areas? In a sense, when one discusses regions $R_1, \ldots, R_4$ as part of some function (such as avoiding threats in chapter 3), the decision to *not* discuss other areas is fairly arbitrary. We could have discussed the roles of regions R_5, R_6, and so on. One of the main reasons we don't is due to the limitations of the tools available to neuroscientists, which are ill-suited to investigating large-scale, distributed systems (although techniques are advancing fast; see chapter 12). In the end, we still don't know much about collective computations involving larger numbers of gray matter components.

Because the concept of *function* is so central to our discussion, chapter 4 is entirely devoted to unpacking the idea. This is important because of the knee-jerk tendency to think in terms of "one area, one function"—for example, the function of the amygdala has to do with emotion or fear. In contrast, chapter 4 describes how a given brain area is always involved in multiple functions, in effect exhibiting a *functional repertoire*. But if so, how should we think of brain areas?

In chapters 5 to 7, I discuss several much-researched regions in relative isolation, providing historical context. This exposition provides some of the basics that will then allow us to delve into their large-scale networks. Chapters 8 to 11 are intended to work as a unit to advance how we need to embrace networks, fully, to understand the brain and behavior at a deeper level. So, as you read chapters 5 to 7, bear in mind the present considerations when a given brain region (say, the striatum) and some of "its functions" are discussed. In chapter 11, we reach the point where we can put things together and see that neural processes and mechanisms are not bound by territorial boundaries. There is no circumscribed place in the brain where, say, "reward" resides—instead, processes and mechanisms related to reward span multiple sectors of the brain. Finally, chapter 12 concludes our journey by revisiting some of the central themes covered in the preceding chapters.

A comment on the word "entangled" in the book title, which conjures multiple interrelated ideas. What I roughly want to convey by using it is *not* something like threads that are mixed together but can be separated if only one has enough time and patience. The meaning I want to convey is closer to "integrated," but single words do not do justice to the general theme permeating the book—for example, cars are highly integrated systems, but are designed with parts with well-defined functions. Instead, the sense of *entangled* that I want to express is one in which brain parts dynamically assemble into coalitions that support complex cognitive-emotional behaviors, coalitions comprised of parts that jointly do their job.

2   Learning a Bit of Anatomy

Without anatomy, we can't learn much about the brain. Knowing where things are and understanding what is connected to what are starting points to deciphering function, so we must familiarize ourselves with the basics of neuroanatomy. Whereas pure memorization plays an important role, understanding some general principles is also essential.

There's no way around it. Anatomy might be dry, but we need it to be able to navigate around the brain. This chapter provides a very brief overview of neuroanatomy that should be helpful to readers unfamiliar with the basics, whereas others may consider skimming the material. Anatomy and function are never far from each other, and some of the discussion below will deal with conceptual issues of understanding the relationship between structure (brain tissue) and function (behavior); the sections on "biology's axiom," "a brief detour into software," and "thinking about networks, not regions" should be of interest to those more familiar with anatomy, too.

When we think of the human brain, the first thing that comes to mind is the cortex—the outer zone of the cerebrum with bumps and grooves (figure 2.1a). When the brain is sliced and appropriately stained to mark the presence of neurons, we see that the cortex is a thin enclosure of densely packed cells (figure 2.1b). The cortex ("bark" in Latin), like the exterior covering of the trunk and branches of a tree, envelops the brain like a rind. In humans and some other mammals, the cortex is not smooth but highly convoluted; if spread like a dough, it would be the size of a large pizza, so the invaginations help pack a larger brain inside the skull. Microscopically, the cortex has a fine layered structure, like a mille-feuille dessert (containing three layers of puff pastry alternating with two layers of pastry cream), of varying cellular complexity. Although at first glance the cortex looks the

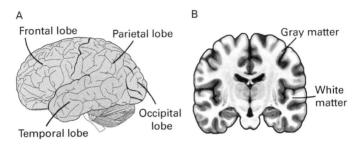

Figure 2.1
Human brain. (a) Side view shows the four major lobes of the brain. The darker
traces are invaginations called sulci. (b) The cortex is the outermost part called "gray
matter," roughly three millimeters in depth. The white parts, called "white matter,"
contain nerve fibers that anatomically link parts of the brain.

same everywhere, in some sectors it may have as few as three and in others
as many as six discernible cell layers (in some parts sublayers are discernible, too, so even more packing can be identified), and the thickness of the
cortex is not the same everywhere but varies between two and three millimeters. But despite important differences in layering and other properties,
the cortex is relatively the same whether you are at the front or back, or top
or bottom, of the brain.

The brain contains two key types of tissue. *Gray matter* contains neurons, which are thought to be the key processing elements of the nervous
system, as well as several other notable cell types that support and protect
the cellular environment (and likely contribute to computations in ways as
yet poorly understood). *White matter* contains nerve fibers, which are the
cell extensions (called axons) of neurons bundled together and that serve
as communication highways both within and between brain regions. Many
of these nerve fibers are enveloped by myelin, a substance that acts as an
electrical insulator and speeds signal conduction along axons—and gives
the white matter its color. Further below, we'll discuss neurons and axons,
as well as the axons' cousins, the dendrites.

A slice through the brain also reveals concentrations of cells that lie
deeper within it and constitute the aptly named subcortex. Whereas the
cortex is essentially a thin sheet of neurons (more precisely, multiple sheets
stacked together) at the outer edge, the subcortex is wholly different. To the
uninitiated, a two-dimensional slice gives little clue as to the underlying

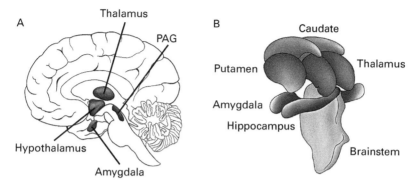

Figure 2.2
Subcortical brain regions. (a) Cut through the middle of the brain illustrates the position of some subcortical regions (PAG is the periaqueductal gray). (b) Rendering of some important structures. The striatum, which is discussed at length throughout the book, has two parts: the caudate and the putamen.

shape of subcortical regions, as this realm is full of oddly shaped masses of tissue (figure 2.2). Here is where neuroanatomist's imaginations have run wild with the challenge of naming structures (possibly creating a modern-day nightmare for medical students prior to exams). Knowledge of Greek and Latin comes in handy because we find structures with names like hippocampus (for its seahorse shape), amygdala (for its almond shape), caudate (for looking like a tail), and even substantia nigra—when creativity fails, name it (the thing/substance) for its color when chemically stained.

Despite the complexity revealed by slicing brain tissue, the brain's organization can be better appreciated if we consider how is forms embryologically. The entirety of the organ originates from a structure called the neural tube, which is literally shaped like a cylinder. At some point during embryonic development, this tube, which is fairly regular at first, creases at three places and bulges so that four compartments can be discerned (figure 2.3). These are the forebrain (front brain), midbrain (middle brain), hindbrain (back brain), and spinal cord. The last three contain no cortex; the first contains both the cortex and several subcortical structures, which we'll learn about later. If it seems confusing to have both cortex and subcortex in the forebrain, remember that this is not arbitrary. Embryologically, they both originate from the segment of the neural tube that differentiates into the forebrain. As the embryo changes in shape, names are needed to keep track

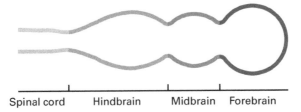

Spinal cord   Hindbrain   Midbrain   Forebrain

Figure 2.3
Embryonic brain sectors marked to indicate the portions that give rise to segments of
the adult brain. For example, the rightmost sector gives rise to all forebrain regions,
including the entire cortex, and all subcortical structures above the midbrain.

of its multiple parts. And as these shapes evolve into a mature state, the
forebrain gives rise to both the cortex externally and multiple subcortical,
inner structures.

The Outer Blanket of the Cerebrum: Cortex

Although the cortex has a fairly regular structure throughout, it is large
enough that we need to subdivide it to be able to orient ourselves (see
figure 2.1). Its main parts are named after skull bones to which they are adjacent: *occipital*, in the back of the skull; *temporal*, near the temples and on
the side of the skull; *parietal*, at the side and top toward the back of the
skull; and *frontal*, around the front part of the skull. Two other parts of
the cortex are not visible from the outside and need to be seen from the
inside to reveal themselves: the *cingulate* (see figure 6.1), which lies along
the middle part of each hemisphere (the brain is made up of two halves, or
hemispheres), and the *insula* (see figure 6.6), which is hidden by the "lid"
of the frontal and parietal cortices (the name insula comes from "island,"
and indeed this part of the cortex is somewhat like an island off the parietal, temporal, and frontal cortices). Finally, the hippocampus is a cortical
structure with simple lamination located close to the medial wall of the
hemispheres (where they come closest to touching).

Although we often associate the cortex with the human brain, this type
of tissue is present in all mammals. (Curiously, it is observed in parts of the
forebrain of some reptiles, although it is fairly small. And as discussed in
chapter 9, fishes, amphibians, and birds have brain parts that are related
to those containing the cortex in mammals.) In a basic sense, what makes

the cortex "cortex" is that it contains a laminated pattern. The six-layer cortex found only in mammals appears to be related to the three-layer cortex found in some reptiles. Indeed, it has been suggested that common elements in three- and six-layer cortex are much like the common set of bones of the basic vertebrate skeleton: "Just as the hand has been adapted from forelimb digits by evolutionary pressures, so have the circuit elements of the basic cortical microcircuit become adapted during cortical evolution" (Shepherd 2011, 44). Such commonalities in traits are called *homologies*, which is to say that they stem from a common evolutionary ancestor. Note, however, that reptiles and mammals diverged more than 300 million years ago! But let's not get ahead of ourselves. We will discuss the evolution of the vertebrate brain in more detail in chapter 9.

Despite the heterogeneity of layering across the human cortex, this tissue type is relatively uniform. If scientists are often classified as "lumpers" or "splitters," it is probably safe to say that most neuroanatomists would fall squarely in the latter camp. In fact, it is not surprising that neuroanatomists have been trying to break up the cortex into smaller parts from the very beginning of neuroscience as a modern discipline (as discussed in chapter 1).

Biology's Axiom

In 1899, upon arriving in Berlin, Cécile and Oskar Vogt established the Neurobiological Laboratory, at first a private institution for the anatomical study of the human brain. Cécile Vogt was one of only two women in the entire Kaiser Wilhelm Institute for Brain Research (which included the Neurobiological Laboratory). In Prussia (with Berlin as its capital since 1701), until around the end of World War I, women were not granted access to regular university education, let alone the possibility to have a scientific career (Cécile obtained her doctoral degree while she was still in France and studied myelination in the cerebral hemispheres). The Vogts collaborated scientifically from 1899 to 1959, an effort that at first depended solely on the earnings from their private medical practice and on the support from the Krupp family, a 400-year-old German dynasty.[1] Two years after the Neurobiological Laboratory was founded, Korbinian Brodmann (chapter 1) joined the group and was encouraged to undertake a systematic study of the cells of the cerebral cortex using sections stained with a new cell-marking method.[2]

Cécile and Oskar Vogt, and Brodmann working separately in their lab, were part of a first wave of anatomists trying to establish a *map* of the cerebral cortex. Neurons are diverse, and several cell classes can be determined based on both shape and size. Researchers used these properties, as well as spatial differences in distribution and density, to define the boundaries between potential sectors. In this manner, Brodmann subdivided the cortex into approximately 50 regions per hemisphere.[3] The Vogts, in contrast, thought that there might be over 200 of them, each with its own distinguishing cytoarchitectonic pattern (that is, cell-related organization). Brodmann's map is the one that caught on and stuck, and today students and researchers alike still refer to cortical parts by invoking his map. Although relatively little was known about the functions of cortical regions at the time, Brodmann believed that his partition identified "organs of the mind"—he was convinced that each cortical area subserved a particular function. Indeed, when he joined the Vogts' laboratory, they had encouraged him to try to understand the organization of the cortex in light of their main thesis: different cytoarchitectonically defined areas are responsible for specific physiological responses and functions.

There is a deep logic to what the Vogts and Brodmann were following. In fact, it is an idea that comes close to being an axiom in biology: Function is tied to structure such that, in the case at hand, parts of the cortex that are structurally different (contain different cell types, cell arrangements, cell density, and so on) carry out different functions. In this manner, they believed they could inform the understanding of how function is implemented from a detailed characterization of the underlying microanatomy. In effect, they were in search of the *functional units* of the cortex. Unlike other organs of the body which have more clear-cut boundaries, the cortex's potential subdivisions are not readily apparent at a macroscopic level. One of the central goals of many neuroanatomists in the first half of the twentieth century was to unravel such "organs" (an objective that persists to this day!). A corollary of this research program was that individual brain regions—say, Brodmann's area 17 in the back of the brain—implemented specialized mechanisms (in this case related to processing visual sensory stimuli). Therefore, it was vital to understand the operation of individual parts as the area was the rightful mechanistic *unit* to understand how the brain works.

Although the brain map produced by the Vogts, with close to 200 areas, was not widely adopted by the scientific community, their approach was clearly superior to that of most of their contemporaries (Brodmann included) because they explicitly compared cellular and functional data. They performed electrophysiological studies in patients and monkeys (they also studied cats) and compared the independently achieved architectonic and functional results in both species to clarify the operation of structurally defined areas (Vogt and Vogt 1926; Amunts and Zilles 2015). Of course, electrophysiological methods of the time were rather crude and amounted to using low-intensity electrical stimulation to observe what behaviors were produced (see chapter 5). Whereas the techniques available severely limited what could be learned (though they produced many pioneering observations), theirs was an extremely advanced conceptual framework that continues to inspire neuroscience today.

A Brief Detour into Software

Let's reconsider "biology's axiom"—different structure implies different function—in the context of concepts outside biology. The development of the computer in the 1940 and 1950s led to important new insights, including those of "software" and "hardware." Software stands for a set of basic instructions or commands that, together, determine how some algorithm is computed (say, factorizing a non-prime number like 2010, which can be written as $2 \times 3 \times 5 \times 67$). In a stunning paper published in 1936, Alan Turing devised an imaginary machine that was capable of calculating *anything* that can be algorithmically computed! This was, of course, before any hardware computer was ever constructed. The first computers were built in the 1940s (although most of us would not recognize them as such in a museum), including the Colossus, which was created to help British code breakers (Turing included) read encrypted German messages during World War II.

The ideas by Turing, as well as those by the mathematicians Alonzo Church, Kurt Gödel (among the all-time most significant logicians), and John von Neumann (famous for designing the basic logic architecture of modern computers), among many others, had a profound effect on philosophers and scientists trying to understand the notion of "computation" in both natural systems (including the nervous system) and artificial ones. An influential framework to emerge in the new field called "philosophy of mind" was that of *functionalism*, which asserts that mental states are identified by their functional role—not by how they are physically implemented.

Thus, a "mind" can be instantiated by various physical systems, possibly even computers, as long as they carry out appropriate computations. According to functionalism, the human brain is one of possibly many physical devices capable of implementing mental functions. In theory, at least.

If some of this feels like armchair philosophy to you, well, it is. Until it isn't. To this day, neuroscience faces these very questions. To what extent do different structural properties (say, neurons of different shapes) affect the functions carried out? Under what circumstances do different organizations of structure lead to similar computations? And so on. As an example, the avian brain is organized in some ways that are rather different from that of mammals. Although a cortex is not found in birds, the dorsal forebrain of the two groups of vertebrates appears to follow similar computational principles that are implemented differently (Dugas-Ford and Ragsdale 2015).

Before proceeding, let's define a few terms of orientation that allow us to navigate up, down, front, and back in the brain. Convenient terms for up and down are "superior" and "inferior" but "dorsal" and "ventral" are used correspondingly, too. The back of the brain is arbitrarily defined as "posterior," so the front is "anterior." Sometimes these terms need to be used carefully because, whereas the human brain is vertically oriented with respect to the main body axis, in other species the body axis is horizontal (think of a fish). But we won't worry about that too much in the book.

The Great Cell Masses: Subcortex

Describing subcortical structures would fill an entire book—and that's a massive understatement. That's not only because there are already quite a few books written about them, but because each structure is pretty complex and heterogeneous. For example, the amygdala, a region that is popular enough that most readers will have encountered it a few times in the popular media, extends no more than 10 millimeters (mm) along its longest axis and 6 mm in the orthogonal one (it is shaped more or less like an almond). Yet, as mentioned in chapter 1, it has more than a dozen subparts (they are called "subnuclei") that are structurally different (given varied neuronal types and patterns of input-output connections) and possibly even more, depending on how it is partitioned.

As we know, the forebrain contains both cortex and subcortex. The subcortical part is located at the base, toward the middle (see figures 2.1b and 2.2).

Many prominent subcortical structures are found there, including the thalamus, hypothalamus, amygdala, and striatum (for the latter, see figure 5.10). At times, the hippocampus is listed as a subcortical structure, but technically it isn't part of the subcortex (even neuroscientists slip here because of its close association with other subcortical areas and simple laminar structure.)

Among the most important subcortical regions of the forebrain is the thalamus (figure 2.2), which lies at the "inner chamber" of the brain (from the Greek *thalamos*, or "chamber"). The term was used by the Greek physician Galen in *De Usu Partium* by way of comparing the human brain with the ground plan of a Greek house, with the bridal chamber at its heart (whereas the name "thalamus" is still used, Galen was probably referring to what's called the third ventricle today[4]). For vision, audition, somatosensation, and taste, individual pathways carrying signals from the sensory periphery pass through the thalamus before reaching the respective cortical areas. For instance, fibers (that is, bundles of axons) leaving the retina of the eye are directed to a part of the thalamus that is connected with the visual cortex in the back of the brain (in the occipital cortex); the part receiving thalamic projections is called primary visual cortex or area V1, for "visual area one." Likewise, fibers leaving the inner ear, after some stops along the way, reach a part of the thalamus that is connected with the auditory cortex (in the temporal cortex); analogously, this part of the brain is called primary auditory cortex or area A1, for "auditory area one." But the thalamus is much more than a simple "relay station" for sensory information reaching the cortex. Anatomists subdivide it into more than 10 subregions with complex connectivity patterns with both the cortex and a very rich array of subcortical regions. Indeed, in later chapters, we will discuss how the thalamus is critically involved in cortical-subcortical loops that play essential computational roles.

Adjacent to the thalamus, we find the striatum, so named given its striped or furrowed appearance. Macroscopically, it contains a few subdivisions, including the caudate and putamen (figures 2.2 and 2.4). A remarkable property of the striatum is that, with the exception of the primary visual cortex, *all* of the cortex projects to it, from sensory regions with simple responses to frontal areas that participate in abstract processes. The striatum projects to subcortical regions, among others, that have a direct impact on motor actions. Indeed, historically, the striatum and adjacent structures forming what is called the basal ganglia (plural for ganglion or cell mass) have been understood as a "motor system." As stated in chapter 1, as early as 1664, Thomas Willis described the striatum, noting degeneration of this structure in patients who suffered from severe paralysis, an observation that led him to link it with body movements (he believed the striatum contained channels for the flow of spirits controlling the muscles).[5] Throughout the book, we will discuss how the striatum, in particular, and the basal ganglia, more generally, are involved in much more than motor functions.

In humans, below the forebrain, the central nervous system extends downward into the midbrain, hindbrain, and spinal cord. The brainstem frequently refers to the large collection of structures in the midbrain and hindbrain (figure 2.2), although the usage is not always consistent across authors. As the name suggests, the overall arrangement resembles a stem on top of which the rest of the brain stands. Given that the brainstem is relatively large, it is typically subdivided into three sectors: the midbrain itself, in addition to the pons and medulla in the hindbrain. These three sectors are quite complex and far from homogenous, and they contain dozens of small zones or areas, each of which participates in multiple functions. The brainstem is the home of many circuits essential for basic processes, such as breathing and controlling heart rate—in short, the regulation of life. In fact, damage to the upper brainstem can cause coma and the so-called persistent vegetative state of partial arousal but not true awareness, in which patients can open their eyelids occasionally and exhibit sleep-wake cycles but completely lack cognitive function. Stroke affecting the brainstem can also cause "locked-in syndrome," in which the patient is completely paralyzed but remains conscious, a devastating condition hauntingly described from the first-person perspective in the book *The Diving Bell and the Butterfly* (popularized as a movie, too).

The Specialized Cells of the Nervous System: Neurons

The human body has over 200 cell types that make up our tissues. An adult male human brain has approximately 86 billion neurons (16 billion of which are in the cerebral cortex), the cell type that is believed to be responsible for most of its unique functions.[6] (The current estimate is a downward revision from the popular 100 billion figure that appears to have been a guesstimate; the number 86 billion may be revised, too, given that it is based on a very small sample of brains, all male at that.) Neurons themselves are a diverse group of cells, including cells specialized to capture sensory information (such as light sensing cells in the eyes) and motor neurons that innervate particular appendages (such as feet) and lead to muscle movements. A typical neuron has two main distinguishable parts: a region that contains the cell nucleus and a set of thin, tube-like radiations that extend outwardly. The central part is called the cell body, or soma. The radiating tubes are of two types: axons and dendrites (figure 2.4).

The cell body usually gives rise to a single axon, which can extend over great distances. Some of the longest ones can exceed a meter, such as the ones from the lower back to the big toe (the part of the nervous system outside the cranium is called the "peripheral nervous system"). Because neurons extend long distances, it was suspected early on that they acted as "wires" that carry output signals. Dendrites, on the other hand, are quite short and rarely longer than two millimeters. Because dendrites come in contact with many axons, they were suggested to act as neuronal "antennae" and contribute to collecting incoming signals.

One of the main ways that neurons communicate with one another is through *action potentials*, also called spikes or nerve impulses. An action potential is only triggered if the electrical voltage crosses a threshold value, at which point it is generated in an all-or-none fashion. Thus, the electrical

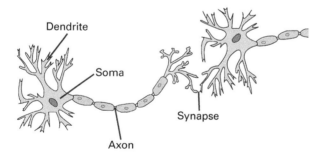

Figure 2.4
Schematic diagram of a neuron. Radiating away from the cell body (also called soma), we see an axon, the component that typically conveys signals to other neurons (this axon is enveloped by a myelin sheath that speeds up impulse transmission). The other extensions are dendrites, which are parts that typically receive inputs from other neurons. The synapse refers to the end of the axon, the narrow space between neurons, and the dendritic contact on the postsynaptic cell. The synapse is where chemical communication between cells takes place.

signal generated at the base of the axon travels along its entire extent at a constant intensity as it propagates. In this manner, the nerve impulse functions as a binary on/off signal, but the frequency and pattern of action potentials provide key information that is communicated between neurons.

(Action potentials are measured by inserting microelectrodes into the extracellular space and measuring electrical voltage. Functional MRI, a technique we'll also discuss in the book, does not measure electrical signals but instead oxygen consumption at a small patch of tissue. But because neuronal activity is costly metabolically, and thus consumes oxygen, functional MRI measures a proxy for neuronal electrical activity. Whereas the technique is very useful in studying brain function, it is important to remember that it provides an indirect measure of the electrical signals that neuroscientists are most interested in.)

How does information pass between neurons? A chief mode of transmission is through *chemical synapses*. Synapses are sites of quasi-contact between neurons, often between a cell's axon and another cell's dendrite. When nerve impulses reach the end of an axon, they cause the release of specific chemicals at the synaptic cleft (the narrow space between cells, 20 to 50 nanometers wide) called *neurotransmitters*. A neurotransmitter released from the presynaptic cell acts on the postsynaptic cell by altering the latter's cell membrane permeability properties, producing excitation or inhibition of the postsynaptic cell (due to the inflow or outflow of ions). When the postsynaptic cell is sufficiently excited, it will generate an action potential, thereby propagating an electrical signal that influences other neurons downstream. This cascade of firing, reverberating across the brain, is at the core of all mental activity! But remember that communication is not only electrical, like a set of electrical cables passing their signals along. It is electrochemical—and arranged in a way that multiple signals can converge and be integrated to lead to further action potentials.

What prevents the brain from going into uncontrolled firing, in effect creating an uncontrollable electrical storm? Indeed, if unchecked, excitation can lead to seizures, from relatively mild to extreme ones. So-called tonic-clonic seizures (formerly known as "grand mal" seizures) can be the most frightening to observe. Typically, the person suffering from such a seizure initially stiffens and loses consciousness, thus falling to the ground. During the second phase, the muscles may begin to spasm and jerk. This terrifying experience mercifully lasts only a few minutes, although it can certainly seem like forever if one is helplessly watching it. It's perhaps not entirely surprising, though utterly tragic, that in Europe of the Middle Ages, epilepsy was confused with witchcraft, especially when accompanied by tremors, convulsions, or loss of consciousness. But what prevents undampened excitation? Neurons influence each other not only in an excitatory fashion but also through inhibition. In the latter case, when a neuron fires, it makes the neurons connected to it *less* likely to generate an action potential.

Neurotransmitters are very diverse (around 100 different molecules have been cataloged), but approximately 10 of them do most of the heavy lifting. They go by names such as dopamine, serotonin, acetylcholine, histamine, and so on, some of which are even household names. For example, antidepressants like Prozac and other variants (fluoxetine, paroxetine, etc.) act on serotonin neurotransmission. These medications are in fact called "selective serotonin reuptake inhibitors" (SSRIs) and lead to an *increased* effect of serotonin on the postsynaptic cell. More generally, alcohol and drugs, including "recreational drugs" like cannabis and hashish, all affect neurotransmission, thereby leading to altered states of consciousness that modify perceptions and feelings. If you thought chemistry was boring, think again.

It is quite humbling that we don't really know how SSRIs work; the mechanisms of action are not well understood. Like many medical treatments, they were discovered by accident, and physicians prescribe them for depression and anxiety based on clinical experience. Ralph Adolphs and David Anderson go as far as suggesting that "trying to cure these [depressed] patients without understanding how the brain generates an emotion state would be like trying to cure the bubonic plague in the fifteenth century without understanding that bacteria and viruses cause infectious disease" (Adolphs and Anderson 2018, 32). As discussed in chapter 1, neuroscience is "observation rich" but not "mechanism rich." We know rather little.

The Massive Highways System

Gray matter is so important that the other part, white matter, receives short shrift. Gray matter is where all the cellular action takes place, white matter is "just a bunch of cables," or so it goes. Much of the communication in the brain occurs locally—for example, within specific areas in the cortex or between two adjacent areas (say, the amygdala and the hippocampus). In such cases, axons are relatively short. However, another type of

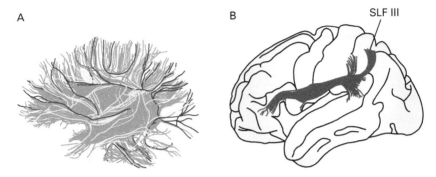

Figure 2.5
White matter. (a) The extensive white matter fibers of the brain interlink brain regions. (b) The fibers are organized in terms of fasciculi (a fasciculus is a bundle of axons), such as the superior longitudinal fasciculus (SLF) III.

connectivity relies on *white matter tracts*—namely, bundles of axons packed together that form a major road system that is essential for signal transmission across the brain (figure 2.5).

If we consider a brain slice (such as in figure 2.1) and mentally remove the outer layer of cortex and other internal clumps that are stained for cell bodies (that is, subcortical areas), it is surprising at first to consider how much remains. All of that is white matter! Anatomists have identified around 20 large tracts that interlink different lobes. Particular tracts differ a little from person to person (slightly thinner/thicker or angled a little differently), but they are found in all typical brains.

Thinking about Networks, Not Regions

White matter is of obvious interest to anatomists and neurologists. When this tissue is compromised, perhaps because of a tumor removal, behavioral deficits are observed. But consideration of white matter has implications that are conceptual in nature and inform one of the central questions occupying neuroscientists: How are functions implemented in the brain?

The dominant theoretical stance in neuroscience has tended to view functions as the product of a particular brain areas—somewhat like a dedicated computer chip that performs specialized computations.[7] As we saw, this idea was very much in line with Brodmann's anatomical research program, and indeed it was part of scientific zeitgeist at the turn of the twentieth century. The existence of fiber tracts was known since the sixteenth century. In *De Humani Corporis Fabrica* (published in 1543), Andreas Vesalius provided a comprehensive account of the corpus callosum and recognized that it links the two halves of the brain. Yet, anatomists like Brodmann who sought to produce "brain maps" paid little attention to the white matter of the brain. But Brodmann's wasn't the only view.

During the last decades of the nineteenth century, at the same time that many researchers were busy studying the impact of lesions to specific parts of the brain, a different school of thought was emerging. The *associationists* argued that, if one observed behavioral change caused by a lesion in a brain area, the deficit could be due to impairment in regions distant from the damaged site. Although magical and spiritual influences were very much in vogue during this period (social gatherings to summon spirits, called séances, were popular at the time), nothing of the sort was being proposed here. Instead, two not mutually exclusive mechanisms were entertained for such "action-at-a-distance" effects: *diaschisis* and *disconnection*. Diaschisis (from the Greek and meaning roughly "shocked throughout") meant that a given region was affected because it was connected with a damaged area, which thereby produced a disruption of the function of the former. The disturbances could be relatively mild but could also be consequential. On the other hand, "disconnection" refers to a situation in which two intact areas are partially or completely disconnected because of an insult to the major tract linking them. Although the two areas remain unperturbed (in contrast to the case of diaschisis), they still may exhibit disturbances of function leading to considerable behavioral alterations. Why? Because their functions depend on their talking to each other. A prime example is the disconnection of the so-called Wernicke's area in parietal cortex and Broca's area in frontal cortex (because of the damage to the tract that interlinks them), regions that play important roles in speech production and language comprehension. What the associationists were hinting at can be viewed as an early incarnation of "network theories." In a nutshell, brain functions are not carried out by single, isolated regions but by coalitions of regions that may be involved in *neural circuits* that are not local—for instance, involving parts of parietal and frontal cortex in the case of speech and language.

Coda

Learning about neuroanatomy can be rather dull. That is in part why it is common to teach students about cortical and subcortical organization by pairing regions with their "main" function, or a small set of functions; say, the hippocampus is important for memory, the prefrontal cortex is important for attention and reasoning, and so on. We will avoid this approach here not because of a possibly better didactic approach but because of the central thesis of the book: Brain areas don't compute specific functions—they are *not* segregated "organs of the mind," as Brodmann put it. The brain is not a modular system that can be understood a region at a time. Instead, we need to unravel how collections of cortical, subcortical, and brainstem regions work together to support complex behaviors. And, as discussed in chapter 9, this is not just the case for the human brain but across all vertebrates—even "simple" ones.

3 The Minimal Brain: Building Simple Defenses and Seeking Rewards

With a system as complex as the brain, where should we start describing it? In this chapter, we describe the idea of a hypothetical "minimal brain" that allows an animal to defend itself and seek rewards, essential components of survival. How do sensations lead to actions through simple sensorimotor circuits? We'll see that action flexibility necessitates uncoupling sensory and motor components. In fact, a brain can be thought of as an entire circuit "in between" sensory and motor cells. This "solution" frees animals from acting simply based on sensory stimulation. Instead, a multitude of factors that encompass emotional and motivational variables are integrated with perception and action to allow successful navigation of the environment.

You are reading this book now, either on paper or on a high-resolution screen, by engaging an incredibly complex visual system. An imaging device, perhaps a new-generation magnetic resonance imaging (MRI) machine, trained on your brain would reveal a large array of visual areas, among many others, that participate in a fine orchestration that takes years to master (think back to elementary school). Now, look around you. Primates, including us, are used to this IMAX, spectacular worldview of detail and color. In primates, when researchers poll the various pieces, the "visual cortex" adds up to roughly a third of the entire cortex.

In the 1880s, experiments with both dogs and monkeys pointed to the occipital cortex as an important territory for vision. In the subsequent decades, the systematic study of clinical cases of patients with blindness, either complete or of one side of the visual field, led to the localization of the visual cortex in the occipital cortex in humans, too. A hundred years after Hermann Munk described his findings about vision in dogs and monkeys, David Hubel and Torsten Wiesel would be awarded the Nobel Prize for Medicine and Physiology in 1981 for their work on the visual properties of neurons in parts of the thalamus and several cortical areas. The centerpiece of their work was the description of how cells in the primary visual cortex (V1) generate their responses, allowing the visual system to respond to contours and boundaries—the basic building blocks of perceiving the shape of objects.

Humans who have lesions in this part of the brain are blind, with the extent of the blindness depending on how much cortex is compromised. Puzzlingly, this is not what George Riddoch, a temporary medical officer in the British army, observed when he examined wounded World War I combatants. Riddoch reported his findings in an article published in 1917, where he described how soldiers who had been blinded by gunshot wounds that had destroyed the visual cortex around the calcarine fissure could still see motion in their "blind" fields, though not much else (see Riddoch 1917).[1] (A fissure, also called "sulcus," is a groove in the cortex; a protrusion in the cortex is called a "gyrus.")

The findings reported by Riddoch and a few others lay dormant for many decades, most likely because they countered the prevailing view that occipital cortex was *necessary* for vision. In 1973, another study reported on the effects of gunshot wounds to the occipital cortex in the back of the head. This time other scientists paid more attention. The patients investigated admitted to no visual experience in the part of the visual field affected by the lesions—to them, they were blind there. Yet, they could move their eyes toward small visual targets presented in the "blind" parts of space if prodded enough by experimentalists. Admittedly, performance was poor, but statistical analysis suggested that it was better than random guessing. How could the patients accomplish this if they did not see the targets?[2] If it crossed your mind that the patients had gained some form of extrasensory perception following their tragic incidents, this is not what was going on. It turns out that there was a second visual system lurking underneath the cortex all along.

Two Small Hills on the Roof

We now know that residual vision is present in persons with a lesion to the primary visual cortex. They may detect the abrupt appearance of objects, movement, and several other visual properties. Indeed, an entire cottage industry of researchers has vigorously studied vestigial visual functions and their implications for understanding the brain, as well as visual consciousness—given that persons with remaining vision frequently are unaware of "seeing." The reasons behind patients' abilities are yet to be completely worked out, but much depends on two small hills in the midbrain at the top of the brainstem. The two structures, one on each side, are called the superior colliculus (where "colliculus" is small hill in Latin).[3] (The superior colliculus is very close to the area called PAG in figure 2.2.)

The retina senses light by transforming energy from photons into electrical signals that leave the eye through a bundle of cables. As discussed in chapter 2, these cables are made of axons, which convey electrical signals between neurons. Action potentials exiting the eye reach the visual cortex in the back of the brain by way of the thalamus. Interestingly, fibers from the retina project to several other places, too, with one group transmitting signals to the superior colliculus at the roof of the midbrain. It is visual processing in this area that is partly responsible for the lingering visual capabilities detected in humans with lesions of the primary visual cortex.

But humans pale in comparison to the skills displayed by tree shrews (they look like small squirrels with pointy noses). Even with the complete elimination of the primary visual cortex, tree shrews exhibit impressive visual behaviors; they avoid obstacles in their path and catch moving pieces of food.[4] Although tree shrews, like all mammals, have visual areas that are cortical, it appears that the balance of contributions to their visual abilities is altered. Humans without a primary visual cortex are blind (but, as noted, some of them demonstrate visual capabilities on careful testing); tree shrews fare much better, in no small part because of the participation of the superior colliculus in their visual behaviors.

Previously, we discussed how the cortex is comprised of layered sheets of neurons and the subcortex is poorly structured. In biology, "rules" always have exceptions, and though part of the brainstem, the superior colliculus is beautifully layered. The number of cell layers varies considerably across species, with some species, such as lizards, having as many as 14 layers (humans and other primates have seven or so layers). Across vertebrates, the top layers, which are the one receiving fibers from the retina, are "visual" and respond to stimuli with short-latency responses. (In vertebrates that are not mammals, the superior colliculus is called the "optic tectum" because it is clearly visible at the "roof" [in Latin, "tectum"] of the midbrain, as shown in figure 9.3. Throughout the book, I'll use "optic tectum" when more clearly referring to nonmammal species.) Retinal projections to the superior colliculus are topographic, meaning that the spatial layout of light hitting the eye (left/right, up/down) and triggering retinal responses is preserved in the colliculus. Cells in the colliculus thus form a *map* of the external visual space, allowing the colliculus to "know" where objects are in the world.

Across vertebrates, the superior colliculus contains intermediate and deep layers, too. The deep layers, in particular, drive action—for instance, head movements in toads and eye movements in primates. To produce movements, signals from the colliculus reach regions in the spinal cord, which themselves control head and eye muscles. Combined, circuits involving the top and bottom layers allow visual inputs to help direct body, head, or eye movements to salient events in the world (figure 3.1).

In all, the superior colliculus, by receiving visual signals from the retina and by driving muscles, accomplishes the essential sensorimotor transformation—whereby sensory stimuli trigger motor responses—required to interface with the world. Input-output arrangements, pretty much necessary for survival, are implemented by the nervous systems of the simplest organisms and, remarkably, can be accomplished by even a single sensorimotor cell (figure 3.2). In this case, the cell, whose body is embedded in the organism, has a *receptor* end that is sensitive to the external environment and an *effector* end that can cause movement of some sort (as simple as some form of contraction). But the single-celled solution is rather inflexible, of course: Pretty much every time the receptor senses something, the effector does its job. The solution, though costly, is to grow more cells in the "middle." And that is what nature did when given a few hundred million years. Indeed, all vertebrates have a superior colliculus—or more generally, a brain. Phrased differently, we can think of the brain, with all its different parts, as evolution's solution to the problem of uncoupling inputs from outputs (figure 3.3). Without this flexibility, animals are bound to perish.

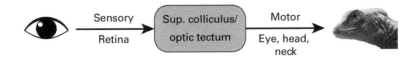

Figure 3.1
The superior colliculus (called the optic tectum in vertebrates other than mammals) receives visual signals from the retina and projects to structures that control movements. The region is often described as a fairly direct sensorimotor interface.

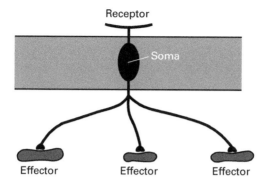

Figure 3.2
Sensorimotor neuron. The middle, elliptical part is the cell body (the soma). The receptor part is above and is sensitive to external stimuli. Effectors can be activated by the axonal ends of the neuron and are capable of generating some kind of tissue motion.

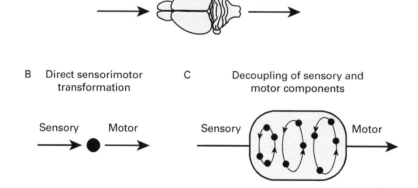

Figure 3.3
Input-output decoupling. (a) The brain solves the problem of decoupling sensory inputs from immediate actions. (b) A fairly direct sensory-motor transformation supports limited and rigid behaviors. (c) Uncoupling input from output provides increased flexibility. The ellipses indicate progressively larger circuits.

Decisions, Decisions

At all times, animals face a crucial three-way decision: Stay the course, move away, or move toward. In rodents, the superior colliculus participates in freezing in place, moving away, and defense-like behaviors.[5] Stimulation of the colliculus can produce a general arousal pattern (which includes large increases in blood pressure and heart rate) and analgesia (that is, processes that ameliorate pain), changes produced during naturally occurring defensive responses.

How is a stimulus classified as harmless, which may or may not be worth investigating further, as opposed to constituting an emergency that requires immediate action? The stimulus's position in the visual field plays an important role here. In small rodents, unexpected movement overhead (much like that of a predator) more likely triggers flight, whereas movement in the lower field (possibly a prey) more commonly elicits approach (figure 3.4). Thus, the superior colliculus could implement a rule much like this: If movement is overhead, flee; otherwise, if movement is in the lower field, consider further exploration. However, simple rules based on

Figure 3.4
Many vertebrates react to stimuli in the upper visual field by fleeing. Stimuli in other parts of the visual field yield other behaviors, including exploration.

elementary stimulus features do not capture the flexibility of rodent behavior (think how hard it is to catch a rat!). For one, rats freeze more frequently to novel stimuli in unfamiliar environments, like an open field. (Freezing is the name given by researchers to a behavior characterized by the absence of overt activity.) Clearly, the context in which a stimulus occurs is paramount. Does the superior colliculus receive additional information that allows it to contribute to behavior in a more malleable manner?

Take the brain of the simplest groups of presently living vertebrates: lampreys, which are water inhabitants with elongated, eel-like bodies, and hagfish, sometimes referred to as slime eels. These animals are important to study because they provide clues about characters that were present in the common ancestor to all vertebrates.

The optic tectum of the lamprey contains five stacks of neurons. As in other vertebrates, the superficial layers receive optic fibers, and the deep layers send outputs that contribute to movements. What types of information do the intermediate layers receive? These layers receive inputs from several sensory sources, including the *lateral line*, a system used by some groups of aquatic vertebrates (including fishes) to detect movement and vibration in the surrounding water. The lateral line plays an important role in maintaining the orientation of the body and in schooling behavior and predation. Notably, in hagfish, the optic tectum receives projections from the hypothalamus which, as discussed below, participates in a wealth of bodily functions (including food intake, thirst, and sexual behaviors).[6] This is really a game changer, as it allows the superior colliculus not only to listen to multiple cues from the external world but also to receive signals from the *internal state* of the body—for example, Are nutrient levels low? Is it time for procreation?—and this is true across all vertebrates, including humans. In this way, the optic tectum's outputs have the potential to reflect multiple variables simultaneously, allowing context sensitivity to emerge as a natural consequence of its wiring and processing (figure 3.5).

The combined information can be used to consider how to react: Stay, approach, or withdraw? As stated, rats freeze in response to novel stimuli in unfamiliar environments more so than in familiar places. Though intuitive, this observation reflects a fundamental principle of brain function—*context sensitivity*. The brain does not simply react to sensory stimuli; instead, incoming data are incorporated into ongoing processing that encompasses the states of the brain and the body, explaining why the exact same stimulus

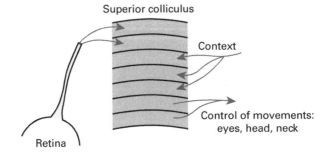

Figure 3.5
Inputs and outputs of the superior colliculus (layers shown schematically). In particular, context signals—for example, from the hypothalamus—allow the region to generate responses that depend on the animal's internal state.

exerts very different effects depending on the situation: in one setting a stimulus may lead to inquisitive approach, in another to moving away.

While the general problem of distinguishing an emergency signal from a neutral one is common to all animals, the details of what counts as an emergency vary markedly between species.[7] This is because adaptive decisions must take into account the relative costs and benefits of orienting toward something ("this is interesting") and escaping ("need to get out of here"). A major benefit of orienting is potentially acquiring better information (the eyes may now be directed at the object of interest); a major cost is loss of time. It's this kind of calculus that varies so much between species. An animal subject to intense predation might have laterally placed eyes that are excellent for panoramic vision but not for seeing details more centrally. In such a case, the gain of an orienting movement to bring a stimulus onto central vision could be relatively slight, whereas the costs of missing a predator would be high. In such animals, therefore, the balance should be tipped in favor of defensive responding, and this expectation is borne out by our experience with rats, rabbits, squirrels, and deer, for instance.

Shoring Up Defenses

The brain is bathed in a colorless fluid also found in the spine called the cerebrospinal fluid that, among other things, provides buoyancy and protection to the brain. The periaqueductal gray, or PAG (pronounced "pee-eyh,-gee"), is an area immediately adjacent to the superior colliculus and that surrounds a channel containing cerebrospinal fluid, hence the name (see figure 2.2). (Although I avoid abbreviations throughout the book, the name periaqueductal gray, also called central gray, is too unwieldy.) The PAG is where outputs from the superior colliculus (and many other brain regions) can be processed into more full-blown defensive programs. In several ways, the PAG can be viewed as an extension of the deep layers of the superior colliculus (Holstege 1991; Brandão et al. 1999).

Unlike the cortex and the superior colliculus, the PAG doesn't have layers. However, the region is not an amorphous bag of cells and seems to be organized based on columns aligned in parallel to the long axis of the brainstem (see figure 2.2). At least two columns of cells help organize defensive behaviors (Bandler and Shipley 1994).[8] In a rat or a cat, excitation of neurons in the "active" column generates behaviors such as facing and backing away or a full-blown flight reaction, and these are very similar to natural actions seen when the animal is threatened or attacked. Excitation of neurons in the "passive" column generates an entirely different response—namely, the cessation of ongoing activity and profound hyporeactivity, with the animal neither orienting nor responding to its environment. This type of freezing behavior is rather similar to that of an animal that has incurred an injury or after defeat in a social encounter (say, being chased by a larger animal). Notably, the PAG generates coordinated actions—that is to say, not simply isolated reactions (like a knee-jerk reflex) but full-blown behaviors. Thus, when engaged by the superior colliculus, the PAG can assist in the production of defensive actions that are beneficial at that point in time.

Seeking Out Rewards

Survival is as much about getting out of the way of danger as it is about keeping the body (think food) and species (think sex) going. When should an animal approach something? Interestingly, the PAG is not all about defense but participates in *appetitive* behaviors, too—that is, those behaviors that increase the likelihood of satisfying specific needs. Sex, for one, is tricky business. Not only are some species hierarchical, with the alpha of the pack having mating privileges, but in many cases females are only receptive during specific periods. Navigate this system poorly and you could end up badly injured.

Lordosis behavior, also called "presenting," is a body posture adopted by many mammals, including rodents, felines, and elephants, that indicates female receptivity to copulation. The body position during lordosis is often crucial to reproduction, as it elevates the hips, thereby facilitating penetration by the penis. Lordosis is commonly seen in female mammals during estrus. Interestingly, the PAG contributes to lordosis behavior, as suggested by impairments in this type of behavior when the structure is lesioned.

The midbrain, where the superior colliculus and the PAG are located, contains other structures that are quite important for appetitive behaviors. A region called the substantia nigra (so named because it appears darker than neighboring areas in chemical preparations) has received a great deal of attention. The superior colliculus has direct connections to the substantia nigra and, importantly, can cause rapid visual activation of neurons there.[9]

The reason the pathway from the superior colliculus to the substantia nigra is particularly noteworthy is that the latter synthesizes and uses the neurotransmitter dopamine. Dopamine, by its turn, plays a significant role in the functions of the striatum, where dopaminergic processing (that is, cellular mechanisms that use dopamine as a key neurotransmitter for neuronal signaling and communication) is important during the processing of novel or salient stimuli. Dopamine has received enormous attention because of its involvement during *reward* processing, including approaching objects previously associated with liked foods. Thus, the pathway from the superior colliculus to the substantia nigra allows the former to participate in appetitive actions rather directly. This is especially the case given extensive connections from the substantia nigra to the striatum and the latter's participation in *motivated behaviors* (for example, "though it might be effortful, I'll move along this path if it brings me closer to obtaining what I want").

Neurotransmitters: A Short Detour

Neurons communicate with one another through chemicals released at their synaptic contacts. Several classes of neurotransmitters have been uncovered, each of which leads to a maze of neurochemical complexity. A peculiarity of several neurotransmitters is that they are synthesized in only a handful of areas. Yet they punch way above their weight because the areas that produce them reach large swaths of the brain through their extensive anatomical connections—what we call "projections systems." The clearest example is norepinephrine, which is contained in only a few small collections of cell bodies in the brainstem. Remarkably, anatomical pathways from these sites go almost everywhere, including cortically and subcortically, enabling this molecule to influence cellular signaling throughout the brain.

Dopamine, too, is manufactured in only a few areas, one of which is the substantia nigra. The dopamine-containing cells there project to the basal ganglia, and this system (substantia nigra plus basal ganglia) has been extensively studied because it is at the root of Parkinson's disease. In the people affected, dopaminergic neurons in the substantia nigra die, causing the symptoms of the disease: most notably, tremors and repetitive movements and difficulty in standing and initiating movements such as walking. Fortunately, the motor impairments can be greatly ameliorated with the administration of L-DOPA, a chemical precursor to dopamine that increases its concentration in the brain.

In the early 1980s, the idea that dopamine is important for motivation and is linked to reward and reinforcement took shape, and since then a tremendous volume of work has shown ways in which this molecule is involved in these processes. How dopamine-related mechanisms are altered in addiction is a question that is actively researched. Most drugs that lead to addiction, including psychostimulants like amphetamine, increase levels of dopamine in the striatum, which can be verified by using an imaging method called positron-emission tomography (PET) that uses small amounts of radioactive drugs to detect specific chemicals in the brain. Studies using this technique show that the participants who display the greatest increase of dopamine after taking drugs are also the ones reporting the most intense "high" or feeling of euphoria (Volkow et al. 2009).

It is not too surprising, therefore, that dopamine is at times treated almost like a "reward molecule." This infelicitous interpretation is common in the general media and in nonspecialty books. Unfortunately, it is also how some neuroscientists speak. But there is no such a thing as a "reward molecule"—the message is not in the molecule.[10] A particular neurotransmitter is involved in multiple functions, and its effect will vary based on the brain region (and circuit) where it operates, including the behavioral context in question. For one, dopamine in the striatum is not exclusively related to motivationally positive events; the processing of negative stimuli involves this molecule, too. Thus, dopamine is not a "reward molecule" for the same reasons we wouldn't call it a "movement molecule" (given the motoric impairments seen in Parkinson's patients).

This tendency to associate one neurotransmitter with one function is a conceptual shortcoming that impedes progress. Consider the involvement of dopamine in the devastating mental disorder of schizophrenia. In 1949, a French surgeon observed that a chemical created as a new type of color dye in the nineteenth century had a markedly calming effect on some surgical patients (just exactly how they decided to administer chemical dyes to patients is itself quite perplexing).[11] Soon afterward, a related dye was found to have beneficial effects on schizophrenics, and in 1954 it was approved in the United States as a treatment for this condition. The chemical, called chlorpromazine, doesn't cure schizophrenia but can attenuate the most severe so-called positive symptoms, such as false-beliefs (like the thought that one's behaviors are being closely monitored and recorded) and disordered thought.

The success of chlorpromazine and other more effective drugs (like haloperidol) led researchers to the "dopamine theory" of schizophrenia (Crow 1980): Drugs that have therapeutic effectiveness (they have antipsychotic effects) *antagonize* dopamine action. According to this framework, dopaminergic projections from the midbrain to the cortex and subcortical structures is overactivated in schizophrenia, dumping too much of this neurotransmitter in the recipient territories. Although current understanding of the role of dopamine in this mental disorder is considerably more nuanced, we can draw the following point from the basic dopamine theory: schizophrenia is not "caused" by dopamine but by the dysregulation of multiple circuits containing dopamine (and many other molecules) in rather complex ways.

Internal Context Signals: Am I Injured?

Processing by the superior colliculus benefits from signals that convey the organism's current state. Chief among the structures that provide this information is the hypothalamus. As the name implies, the hypothalamus is located just below the thalamus (neuroanatomists weren't very imaginative here); it's also just above the brainstem. By the first years of the 1900s, the hypothalamus was identified as an anatomical entity surrounding the third ventricle (ventricles are cavities in the brain that are filled with cerebrospinal fluid). In 1904, Ramon y Cajal (encountered in chapter 2) described several "nuclear formations" (masses of neurons) forming the hypothalamus.

In 1929, Harvey Cushing, considered by many to be the father of modern neurosurgery, described the functions of the hypothalamus in this way: "Here in this well concealed spot, almost to be covered with a thumb nail, lies the very mainspring of primitive existence—vegetative, emotional, reproductive" (as cited by Card, Swanson, and Moore 2003, 795).

Since the 1920s and 1930s, our knowledge of hypothalamic function has greatly expanded. Current understanding concurs with the earlier notion that the area is involved in multiple "basic" life-preserving operations, including complex homeostatic mechanisms, in addition to contributing to neuroendocrine outputs (the endocrine system involves a network of glands that produce and release hormones that regulate many body functions). As we'll cover in chapter 5, the hypothalamus participates in a bewildering array of processes having to do with wakefulness/sleep, hunger, thirst, sex, and defensive behaviors, among others.

By receiving signals from the hypothalamus, the superior colliculus is thus privy to a host of signals about the internal condition of the organism. Sensory inputs can then lead to motor actions in a way that are appropriate for the animal's state. Is it injured, hungry, sleepy?

Extending the Circuit

We started with the superior colliculus, which receives retinal inputs and can guide movements in a fairly direct way, as some of its connections extend down into the spinal cord and from there can influence muscle movements by way of a single additional connection. We then added the PAG, which helps generate defensive and appetitive behaviors. We considered the substantia nigra, which is particularly important for appetitive behaviors, in a manner that is substantially expanded when we incorporate the striatum, too. We saw that the hypothalamus brings a considerable degree of context dependency to the system (figure 3.6).

Put together, these pieces constitute a sort of "minimal brain," with sensory inputs, motor outputs, and parts in between. Remember that the "in between" is how inputs and outputs get decoupled—no one likes to repeat the same thing over and over, and nature will eliminate anything that does. The overall circuit helps an animal answer the critical question, "Stay, approach, or withdraw?" It helps orchestrate approach and withdrawal actions while an animal navigates challenging environments. The combined

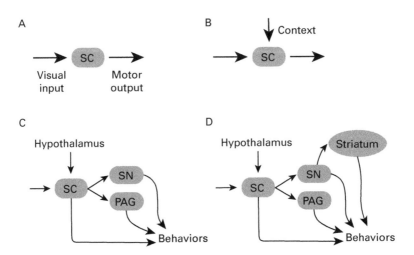

Figure 3.6
Extended superior colliculus circuit (SC). (a) Simple sensorimotor interface, where inputs lead to direct outputs. (b) Context signals can influence the circuit. (c) Connections to the substantia nigra (SN) and periaqueductal gray (PAG). (d) The substantia nigra is also robustly connected with the striatum.

circuit supports an extended behavioral repertoire and frees the animal from responding in the same way every time it receives a specific input.

This mini-brain in the midbrain (and neighboring structures like the hypothalamus and striatum) allowed us to start navigating the central nervous system and considering how some structures contribute to behaviors. To what extent is this exercise of carving out a part of the brain reasonable? Can we separate this mini-brain from the rest, or do we need to consider additional pathways and regions to understand how the mini-brain contributes to behavioral functions? Perhaps the superior colliculus/optic tectum, a key region of this circuit, is sufficiently isolated that such a strategy could work, particularly in "simple" animals. Let's briefly assess this possibility (for a detailed discussion, see chapter 9). In fishes, amphibians, and reptiles, the optic tectum is enormous and wields immense influence (see Striedter 2005). Information processing in these vertebrates heavily includes this area, and its large number of input-output pathways precludes isolating it from the "rest" of the brain in an attempt to understand its functions. Mammals have a forebrain that is rather prominent, including its layered mantle, the cortex. Perhaps in this case the superior colliculus is more isolated and thus a better example of a mini-brain. Au contraire. The mammalian superior colliculus is abundantly interconnected with the remainder of the central nervous system, so much so that one author even suggested that it could help support or even define the "self" (Strehler 1991)—according to the author's logic, the "self" should be located somewhere in the brain that is very well connected anatomically!

This exercise of isolating brain regions and outlining how they contribute to behavioral functions, while useful didactically, is rather unsatisfactory. To some extent, we will continue to resort to it, if only because reading and understanding proceed in a sequential manner. But the reader should remember that we cannot simply point to a brain structure and say that a behavior resides there. Instead, a central thesis of this book is that anatomically *distributed circuits* bring about the behaviors in question. (Even these distributed circuits need to be understood in terms of a fully behaving animal immersed in a broader context.) Bearing this in mind, a potential strategy to appreciate how brain regions contribute to functions and behaviors is to consider the location of interest and a gradually expanding circle of areas to which it is connected. The question is, then: How does this region that we care about help carry out mechanisms of interest *in combination with other regions*?

Before we are able to tackle this problem more directly, we have to build our vocabulary further and familiarize ourselves with many cortical and subcortical areas that have been implicated in domains that a standard textbook would classify as "perception," "cognition," "emotion," and so on. In fact, even before we do that, we need to delve deeper into an issue neuroscientists are faced with front and center: How should we think of individual areas of the brain and their contributions to behavior?

4   What Do Brain Areas Do?

In the past chapters, we encountered several brain areas and some of their functions. But what exactly is an area? Are brain areas more or less like discrete computational units that have well-defined functions? Because this question is so central to our understanding of the brain, we need to slow down here and discuss it in greater depth. We will see that the idea of "one area, one function" largely hinges on assuming a brain organization that is *modular* in essence. In contrast, if a given brain area is always involved in multiple functions, as advocated throughout the book, how should we revise our thinking?

Tan, Tan

When Tan was admitted to the hospital at the age of 21, he had lost the use of speech for some time.[1] He could no longer pronounce more than a single syllable. Whenever a question was asked, he would always reply "tan, tan," accompanied by varied expressive gestures. In fact, throughout the hospital, he was known only by his nickname, Tan. Despite his impediment, at the time of admission, he was perfectly able-bodied and intelligent and appeared to comprehend almost everything that was said to him. (In Brazilian Portuguese, "tan-tan" is colloquially, and pejoratively, used to denote someone who is "crazy." I wonder if the origin has something to do with Tan's predicament.)

In April 12, 1861, about 10 years after his initial admission, and rapidly deteriorating in health, Tan was seen by Paul Broca, a surgeon with an unusual background—he was one of the founders of the field of anthropology in France. Five days later, Tan would die of a severe case of gangrene. The brain was removed and preserved in a fixation fluid that made the tissue harden with time. Based on his examination, Broca concluded that Tan's speech deficit was due to a lesion of the left frontal lobe. Broca concluded his short case report published in the *Bulletin de la Société Anthropologique*, a mere page and a half in length, with the following momentous conclusion: "All this permits, however, the belief that, in the present case, the lesion of the frontal lobe was the cause of the loss of speech." More than 150 years later, Broca's report is the most important paper in the history of brain function localization.

Discovering the Function of Brain Areas

Historically, lesions have played a major role in trying to infer the function of brain subparts. Two types of lesion have been considered: in humans, naturally occurring damage from tumors and vascular accidents; in animals, more precisely delineated lesions produced surgically. Broca's paper, capitalizing on the first type of injury, catapulted forward the idea that a mental function can indeed be *localized*. At the time, very little was known about how the convoluted mass of gray and white matter inside the head supports mental faculties. Consider that Broca's observations took place not long after the heyday of the much-maligned phrenology movement espoused by Franz Gall and his disciples, which was particularly influential between 1810 and 1840. Phrenologists would observe and feel the skull of individuals to determine their psychological propensities such as "philoprogenitiveness" (that is, the love of offspring or children in general), which was located centrally at the back of the head (more or less where we now know the visual cortex to be!).

An early series of lesion studies in animals was performed by Eduard Hitzig and Gustav Fritsch on dogs (published in 1870).[2] Hitzig was a psychiatrist interested in the potential applications of weak electrical currents to ameliorate certain medical conditions. By the mid-1860s, he had developed an apparatus to deliver electrical stimulation to human patients and observed that current applied to the back of a patient's head reliably elicited eye movements, prompting him to investigate the use of the technique (also called galvanization) further. Hitzig thus invited Fritsch, an anatomist, to join him in studies to be conducted in dogs. Their most famous experiment was performed not in a well-equipped university laboratory but on a dressing table in a bedroom of Hitzig's house in Berlin. Initially, they electrically stimulated the canine cortex with weak electrical currents. By systematically varying the site that was excited, they uncovered locations that elicited muscular responses of the forepaw, hindpaw, face, and neck (in all cases, on the side of the body opposite to the stimulation, also called the contralateral side). With a scalpel, they then removed the area that led to, say, forepaw movement upon galvanization. Although this did not abolish all movement from the contralateral paw, movement was impaired, and abnormal postures were observed. Notably, sensation appeared to be normal, as the animals' responses to stimuli were unaltered. That is to say, the observed deficit caused by the lesion was relatively *selective* for motor production (and not sensory perception), and it was even linked to a specific body part.

The importance of the study by Hitzig and Fritsch, like the observations by Broca, cannot be overestimated. This was as much due to their results as to their *conclusions* based on combining electrical stimulation and lesions. As immodestly stated by them: "Some psychological functions, and perhaps all of them, in order to enter matter or originate from it, need circumscribed centers of the cortex." That is to say, according to them, the cortex contained processing *centers*. Hitzig and Fritsch therefore suggested that it would be worthwhile for researchers to search for areas concerned with sensation and even regions involved with intelligence. The time was ripe to explore the locations where mental functions *reside*.

Dissociating Mental Functions

It is definitely possible to study the brain at its most elementary sense: mechanisms of neuronal spike generation and signal propagation along axons; molecular mechanisms along the gap, or synapse, between two neurons; and so on. But, often one studies brain mechanisms, even the most basic ones, to understand the neural basis of mental functions—seeing a sunset, hearing a screech, speaking a sentence, remembering a childhood memory, feeling uncertain about the future.

A chief goal in the sciences of the mind and brain is to explicitly unravel the *functional architecture* of the mind: to identify and characterize the *mental* processes underlying behavior. But mental processes are not directly observable. Rather, their existence must be inferred from external manifestations—what we call overt behaviors. Insight into mental functions can be gained based on the by-products of brain damage on carefully chosen tasks—much like the approach of Broca and of Hitzig and Fritsch. In broad terms, the existence and general contours of mental processes are then inferred from the manner in which task performance changes from manipulation to manipulation, involving different levels of an experimental variable and different forms of brain damage. In trying to delineate mental processes, lesions are extremely valuable even to cognitive scientists not inherently interested in the brain. The reason is that disturbances in behavioral performance when the brain is damaged can inform us about the organization of the *mind*. For example, to what extent is the processing of verbs and nouns separate? If one uncovered damage that affects the processing of words used as verbs (say, "he judges") more than of the same words used as nouns (say, "the judges"), this would be extremely valuable in outlining the organization of the mental processes in question—what was called the functional architecture above.

To infer the existence of *separate* mental processes, researchers rely on the logic of *dissociations*.[3] Consider, first, a *single* dissociation. Let A and B be two tasks (say, one involving verbs, another involving nouns) and let m be a "manipulation." A single dissociation is observed if m affects performance on A but not on B. The "manipulation" could correspond to a lesion of a region, and a dissociation would be established if the damage impaired performance of A but not B. In all, a dissociation invites the inference that there is an underlying mental function required by A but not by B.

The logic of dissociation is central to neuroscience and has long been used to *localize* mental functions. The reasoning is analogous to what one would adopt to reverse-engineer a human-made device—say, remove the pistons in a car to try to "discover" that they are a key element in the combustion process that powers standard automobiles. But single dissociations are inferentially weak, and although they were frequently employed in research in the first decades of the twentieth century (and in many ways after that), in the 1950s investigators started to question the single dissociation's application. For example, in some cases it may well be that *general* deficits following a lesion could explain the pattern of results; perhaps the lesion impairs most tasks that are difficult, and task A happens to be harder than task B.

Evidently, better experimental design and methodology, with careful choice of tasks A and B (and the dimensions along which A and B are matched), ameliorate the problems with guessing function. This is simply good experimental design, which is the cornerstone of solid experimental science. After all, if changes are observed in one experimental condition, the question is always, "Relative to what?" Control conditions are fundamental in drawing reasonable inferences. Nevertheless, in trying to establish the anatomical underpinnings of mental functions, the single dissociation strategy is simply too weak. The inconclusiveness of the methodology motivates the *double* dissociation logic (figure 4.1). A single dissociation is observed if region 1 affects performance on task *A* but not on task *B*. A double dissociation is observed if, in addition, region 2 affects performance on *B* but not on *A*. Both single and double dissociations indicate that there is an underlying mental function required by *A* but not by *B*. In addition, a double dissociation invites the converse inference—namely, that there is an underlying mental function required by task *B* but not by task *A*. In addition, it is surmised that brain regions 1 and 2 carry out functions *that* are relatively *isolable* from each other. The power of the double dissociation logic lies in its specificity: lesions of regions 1 and 2 do not simply cause a series of impairments; instead, they impair *circumscribed* mental functions.

What Do Double Dissociations Tell Us?

In humans, before the advent of modern neuroimaging techniques such as functional magnetic resonance imaging (MRI), researchers tried to uncover function by studying impaired and preserved abilities in brain-damaged patients. And the most powerful weapon in the neuropsychologist's armamentarium was the double dissociation approach just described.

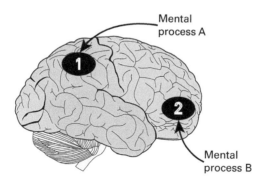

Figure 4.1
The double dissociation logic. If mental process A is affected by a lesion to region 1 but not region 2, and vice versa for mental process B, we say that they are doubly dissociated. To many neuroscientists (but not all!) this pattern suggests that areas A and B are functionally specialized.

Here's a specific example. Although neurologically intact persons have no difficulty pronouncing written words regardless of whether they have a concrete meaning (say, "table") or an abstract meaning (say, "truth"), after a severe left-hemisphere stroke, patient PW correctly pronounced 67 percent of concrete words but only 13 percent of abstract words.[4]

The finding that concrete and abstract words were differentially susceptible to damage suggested that they are represented separately in neural tissue. An alternative view, however, is that PW's brain damage affected concrete and abstract representations equally, but that abstract words were more impaired because they are inherently more difficult to pronounce. On this latter account, one would not expect to see the opposite relationship: better reading of abstract than concrete words following brain damage. This is exactly what the neuropsychologist Elizabeth Warrington observed in patient CAV who had a left-hemisphere tumor: He read correctly 36 percent of concrete words but 55 percent of abstract words. Together, patients PW and CAV exemplify a double dissociation of concrete and abstract word reading. (Although both lesions were in the brain's left hemisphere, they were assumed to compromise different parts of the cortex. Moreover, the behavioral impairments of PW and CAV were observed in other patients, too.)

Double dissociations among brain-damaged patients, as well as animals with focal lesions, have been identified for many pairs of tasks, spanning perception, action, emotion, and motivation. To exemplify the typical interpretation of this form of relationship, consider the conclusion by Warrington, herself a hugely influential scientist, that "the only plausible interpretation of a double dissociation between abstract-word deficit and concrete-word deficit . . . is that the functional and structural organization of semantic representations of words is categorical" (Warrington 1981). That is to say, the semantics of concrete words and those of abstract words must be implemented separately—they rely on different functions *and* are carried out in separate parts of the brain. The reasoning supporting this interpretation dovetails nicely with the view that mental functions rely on a collection of relatively independent processing components or *modules*, each dedicated to performing a particular function, which is a view embraced by many influential researchers. In fact, double dissociations and modularity go together so naturally that the theoretical perspective of modularity has dominated several subfields of neuroscience. As another neuropsychologist, Andrew Ellis, put it: "There can be no argument with the fact of modularity, only about its nature and extent" (Ellis 1987, 402).[5]

What Is Modularity?

So, how isolable are the parts of the brain? Modularity can be conceptualized in multiple ways:[6]

M1: Two parts A and B of a system are defined as modules, if and only if they are separately modifiable.

M2: The process carried out in the subsystem, so modifiable, computes a particular type of input-output mapping.

M3: There exists a decomposition of the system such that the computational interactions *within* subsystems are much more complex than those *between* subsystems.

M4: Subsystems form into complex networks with other subsystems so that each is carrying out only a particular subfunction of a much more complex overall function.

M5: The subsystem needs to be relatively spatially localized in the brain.

Property M1 provides a generic description of the idea, with M2 further specifying that some input-output relationship should be computed by the module in question. But what are modules? Perhaps they are fairly well delimited parts—say, Brodmann's area 17 in the back of the brain corresponding to the primary visual cortex. But acknowledging that modules might involve more than just a single area, property M3 tries to capture a less restrictive notion of modularity, where one can think of subsystems (themselves perhaps composed of more elementary components) that are relatively encapsulated from *other* subsystems. Property M4 goes a step further, admitting that subsystems themselves are fairly complex and might interact with other subsystems. Still, being a subsystem, it should compute an identifiable "elementary" function.

Properties M1 to M4 are general and could apply to any system, natural or human-made. Property M5 is specific to the brain, of course, and is key to how the notion is conceptualized in neuroscience. Unless some version of property M5 holds, the system would not be recognized as modular. For example, it is possible to imagine a brain whose functions follow properties

M1 to M4 but is physically implemented in a spatially distributed fashion (imagine an artificial brain yet to be produced; incidentally, philosophers and cognitive scientists love to concoct all sorts of such challenging possibilities). In such case, the system would be *functionally* modular but not in terms of how it is instantiated in a *physical* medium.

It is useful to recast modularity in terms of *decomposability* (chapter 1). A decomposable system is one in which each subsystem does its job independently of the others. In contrast, in a non-decomposable system, the components are so interrelated as to defy attempts to break them up. What kind of system is the brain? Where along the spectrum of decomposable to non-decomposable does it reside?

This question is not idle armchair musing. It is at the core of our strategy to investigate the architecture of the mind-brain. Even more so because we must confront head-on the following question: Are there kinds of systems for which a *reductionistic analysis*—that is, one in terms of simpler subcomponents—would fail (Bechtel and Richardson 2010)? Here, reductionism means the type of approach central to science, in which an organization of greater complexity is understood in terms of the contributions of its subparts, which when put together give rise to the behavior of the broader system.

Consider the case of an object in which the components, perhaps simple computational elements not unlike neurons, do not perform operations that are (too) distinct from one another and for which the *interactions* between elements within the system are chiefly responsible for generating its behavior. To add to the difficulties, imagine a scenario in which the interactions between components are nonlinear, where, say, more of an input does not necessarily translate into more of an output (see chapter 8). In such cases, I contend that insurmountable difficulties arise in trying to unravel the object's working by reducing it to that of putative subcomponents.

To be sure, posing the question in this manner may sound counterintuitive to readers (and scientists alike) accustomed to successes of mechanistic analyses—the very bedrock of science. Indeed, reductionism is the declared philosophy of most scientists. Reduce everything to the smallest parts, determine their properties, and you explain the whole system. As developed throughout the book, I believe such an approach provides at best an impoverished description of brain function, as most of the explanatory work needs to be done at the level of *interactions*. Unfortunately, neuroscience as a discipline is all too reductionist. But rejecting the philosophy of reductionism is not an attack on scientific analysis (that is, the decomposition into parts and their analysis) and a concomitant embrace of some ill-specified holism. As stated by Ernst Mayr, sometimes hailed as the Darwin of the twentieth century, "No complex system can be understood except through careful analysis; however, the interactions of the components must be considered as much as the properties of the isolated components" (Mayr 2004, 34). We will have a lot more to say about these issues in chapter 8 when discussing *complex systems*.

One Area, One Function?

Let's go back to brain areas and consider, once more, their relationship to mental processes. We'll start with the simplest formulation—namely, by assuming a one-to-one mapping between an area and its function. (We are assuming for the moment that we can come up with, and agree on, a set of criteria that defines what an area is. Maybe it's what Brodmann defined early in the twentieth century, or perhaps it is as defined in the recent proposal discussed in chapter 1. For example, we could say that the function of the primary visual cortex is visual perception, or perhaps a more basic visual mechanism, such as detecting "edges" (sharp light-to-dark transitions) in images. The same type of description can be applied to other sensory (auditory, olfactory, and so on) and motor areas of the brain. This exercise becomes considerably less straightforward for brain areas that are not sensory or motor, as their workings become much more difficult to determine and describe. Nevertheless, in theory, we can imagine extending the idea to all parts of the brain. The result of this endeavor would be a list of area-function pairs: $L = \{(A_1, F_1), (A_2, F_2), \ldots, (A_n, F_n)\}$, where areas A implement functions F.

To date, no such list has been systematically generated. However, current knowledge indicates that this strategy would *not* yield a simple area-function list. What may start as a simple (A_1, F_1) pair, as research progresses, gradually is revised and grows to include a list of functions, such that area A_1 participates in a series of functions $F_1, F_2, \ldots, F_k$. From initially proposing that the area implements a specific function, as additional studies accumulate, we come to see that the area *participates* in multiple functions. In other words, from a basic one-to-one $A_1 \rightarrow F_1$ mapping the pictures evolves to a one-to-*many* mapping: $A_1 \rightarrow \{F_1, F_2, \ldots, F_k\}$ (figure 4.2a and 4.2b).

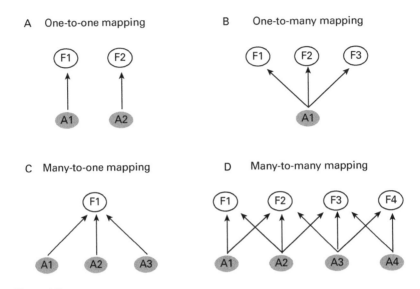

Figure 4.2
Structure-function mapping in the brain. (a) One area, A, might be involved in a single function, F, such as when people suggest that the amygdala is specialized for processing fear. (b) But we know that the amygdala carries out multiple functions. (c) From the standpoint of functions (such as aversive processing), multiple areas may be able to carry it out. (d) Elements of biological systems, like areas of the brain, exhibit the most complex mapping of them all: many-to-many.

Consider this example: Starting in the 1930s, lesion studies in monkeys suggested that the prefrontal cortex implements "working memory," such as the ability to keep in mind a phone number for several seconds before dialing it. As research focusing on this part of the brain ramped up, the list of functions grew to include many cognitive operations, and the prefrontal cortex became central to our understanding of what is called *executive function* (see chapter 7). In fact, today, the list is not limited to cognitive processes but includes contributions to emotion and motivation. The prefrontal cortex is thus multifaceted. One may object that this sector is "too large" and that it naturally would be expected to participate in multiple processes. While this is a valid interjection, the argument holds for "small areas," too. For example, take the amygdala, a region often associated with handling negative or aversive information. However, the amygdala also participates in the processing of appetitive items (and this multifunctionality applies even to amygdala subnuclei).

Let's consider the structure-function ($A \rightarrow F$) mapping further from the perspective of the mental functions: Where in the brain is a given function F carried out? In experiments with functional MRI, tasks that impose cognitive challenges engage multiple areas of the frontal and parietal cortex. Examples are tasks requiring participants to selectively pay attention to certain stimuli among many and answer questions about the ones that are relevant (in a screen containing blue and red objects, are there more rectangles or circles that are blue?). These regions are important for paying attention and selecting information that may be further interrogated. Such *attentional control* regions are observed in circumscribed sectors of the frontal and parietal cortex. Thus, multiple individual regions are capable of carrying out a mental function, an instance of a *many*-to-one mapping: $\{A_1$ or $A_2, \ldots,$ or $A_j\} \rightarrow F_1$. (See figure 4.2c.) The explicit use of "or" here indicates that, say, A_1 is capable of implementing F_1, but so are A_2, and so on.[7] Now, together, if brain regions participate in many functions and functions can be carried out by many regions, the ensuing structure-function mapping will be *many*-to-*many* (figure 4.2d). Needless to say, the study of systems with this property will be considerably more challenging than systems with a one-to-one organization. (For a related case, consider a situation where a gene contributes to many traits or physiological processes; conversely, traits or physiological processes depend on large sets of genes.)

Structure-function relationships can be defined at multiple levels, from the precise (for instance, primary visual cortex is concerned with detecting object borders) to the abstract (for instance, primary visual cortex is concerned with visual perception). Accordingly, structure-function relationships will depend on the granularity in question. Some researchers have suggested that, at some level of description, a brain region does *not* have more than one function; at the "proper" one, it will have a single function (Price and Friston 2005). In contrast, a central idea developed in this book is that the one-to-one framework, even if implicitly accepted or adopted by neuroscientists, is an oversimplification that hampers progress in understanding the mind and the brain.

Brain Areas Are Multifaceted

If brain areas don't implement single processes, how should we characterize them? Instead of focusing on a single "summary function," it is better to describe an area's *functional repertoire*: Across a possibly large range of functions, to what extent does an area participate in each of them? No consensus has emerged about how to do this, but below we'll discuss some early results. The basic idea is simple, though. For example, coffee growers around the world think of flavor the same way: as a flavor profile or palette. Brazilian coffee is popular because it is very chocolaty and nutty with light acidity, to mention three attributes.

Research with animals uses electrophysiological recordings to measure neuronal responses to varied stimuli. The work is meticulous and painstaking because, until recently, the vast majority of studies are recorded from a single (or very few) electrode(s) in a single brain area. Setting up a project, a researcher thus decides what processes to investigate at what precise location—for example, probing classical conditioning in the amygdala. Having elected to do so, the electrode is inserted in multiple neighboring sites as the investigator determines the response characteristics of the cells in the area (newer techniques exist where grids of finely spaced electrodes can record from adjacent cells simultaneously; see chapter 12). For some regions, researchers have cataloged cell response properties for decades; considering the broader published literature thus allows them to have a fairly comprehensive view. In particular, the work of mapping cell responses has been the mainstay of perception and action research, given that the stimulus variables of interest can be manipulated systematically; it is easy to precisely change the physical properties of a visual stimulus, for example. In this manner, the visual properties of cells across more than a dozen areas in the occipital and temporal cortex have been studied. And several areas in the parietal and frontal cortex have been explored to determine neuronal responses during the preparation and elicitation of movements.

It is thus possible to summarize the proportions of functional cell types in a brain region.[8] Consider, for example, two brain regions in the visual cortex called V4 (visual area number 4) and MT (found in the middle temporal lobe). Approximately 85 percent of the cells in area MT show preference for the direction that a stimulus is moving (they respond more vigorously to rightward versus leftward motion, say), whereas only 5 percent of the cells in area V4 do so. In contrast, 50 percent of the cells in area V4 show a strong preference to the wavelength of the visual stimulus (related to a stimulus's color), whereas no cells in area MT appear to do so. Finally, 75 percent of the cells in area MT are tuned to the orientation of a visual stimulus (the visual angle between the major elongation of a stimulus and a horizontal line), and 50 percent of the cells in area V4 do so, too. If we call these three properties DS, WS, and OS (for stimulus direction, wavelength, and orientation, respectively), we can summarize an area's responses by the triplet (DS, WS, OS), such that area MT can be described by (0.85, 0, 0.75) and area V4 by (0.05, 0.50, 0.50), as shown in figure 4.3.

This type of summary description can be potentially very rich and immediately shifts the focus from thinking "this region computes X" to "this region participates in multiple processes." At the same time, the approach prompts us to consider several thorny questions. In the example, only three dimensions were used, each of which related to an attribute thought to be relevant—related to computing an object's movement, color, and shape, respectively. But why stop at three features? Sure, we can add properties, but there is no guarantee that we will cover all the "important" ones. In fact, at any given point in time, the attributes more likely reflect what researchers know and likely find interesting. This is one reason the framework becomes increasingly difficult for brain areas that aren't chiefly sensory or motor; whereas sensorimotor attributes may be more intuitive, cognitive, emotional, and motivational dimensions are much less so—in fact, they are constantly debated by researchers! So, what set of properties should we consider for the regions of the prefrontal cortex that are involved in an array of mental processes?

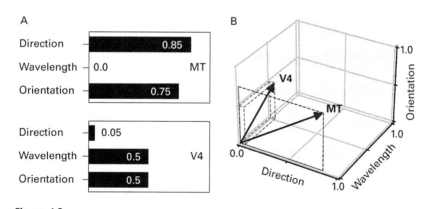

Figure 4.3

Multifunctional description of brain regions. (a) Visual areas MT and V4 can be described in terms of three attributes: direction, wavelength, and orientation. (b) Vector representation of the two areas.

More fundamentally, we would have to know, or have a good way of guessing, the appropriate *space of functions*. Is there a small set of functions that describes all of mentation? Are mental functions like phonemes in a language? English has approximately 42 phonemes, the basic sounds that make up spoken words. Are there 42 functions that define the entire "space" of mental processes? How about 420? Although we don't have answers to these fundamental questions, some form of multifunction, multidimensional description of an area's capabilities is needed. A single-function description is like a straitjacket that needs to be shed. (For readers with a mathematical background, an analogy to basic elements like phonemes is a "basis set" that spans a subspace, like in linear algebra, or "basis functions" that can be used to reconstruct arbitrary signals, like in Fourier or wavelet analysis.)

The multifunction approach can be illustrated by considering human neuroimaging research, including functional MRI. Despite the obvious limitations imposed by studying participants lying on their backs (many people will feel sleepy and may even momentarily doze off, not to mention that we can't ask them to walk around and "produce behaviors"), the ability to probe the brain noninvasively and harmlessly means that we can scrutinize a staggering range of mental processes, from perception and action to problem solving and morality. With the growth of this literature, which accelerated in earnest after the publication in 1992 of the first functional MRI studies, several data repositories have been created that combine the results of thousands of studies in a single place.

In my laboratory, we capitalized on this treasure trove of results to characterize the "functional profile" of regions across the brain. We chose 20 "task domains" suggested to encompass a broad range of mental processes, including those linked to perception, action, emotion, and cognition. By considering the entire database of available published studies, at each brain location, we generated a 20-dimensional functional description indicating the relative degree of engagement of each of the 20 domain attributes (figure 4.4a). Essentially, we counted the number of times an activation was reported in that brain location, noting the task domain in question. For example, a study reporting stronger responses during a language task relative to a control task would count toward the "language" domain at the reported location. We found that brain regions are rather functionally

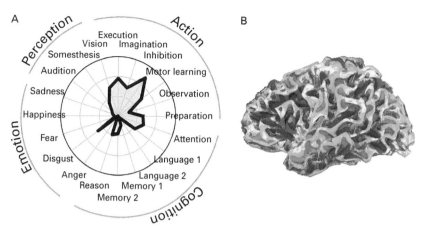

Figure 4.4
Multifunctionality of brain regions. (a) A radial plot shows the functional profile of a sample region. It includes 20 attributes across four classes (perception, action, cognition, emotion). The plot indicates the degree of engagement of the region for each attribute. For example, emotion studies involving disgust-related pictures or words engage the region strongly, but other emotion-related content does not. (b) The color insert shows the distribution of a measure of functional diversity across the cortex (warmer colors indicate higher diversity; cooler colors, less diversity).
Source: Panel B reproduced with permission from Anderson, Kinnison, and Pessoa (2013).

diverse and are engaged by tasks across many domains. But this didn't mean that they respond uniformly; they have preferences, which are at times more pronounced. To understand how multifunctionality varied across the brain, we computed a measure that summarized *functional diversity*. A brain region engaged by tasks across multiple domains would have high diversity, whereas those engaged by tasks in only a few domains would have low diversity. Functional diversity varied across the brain (figure 4.4b), with some brain regions being recruited by a very diverse range of experimental conditions.

The findings summarized in figure 4.4 paint a picture of brain regions as functionally diverse, each with a certain *style of computation*. The goal here was to illustrate the multidimensional approach rather than to present a more definitive picture. For one, conclusions were entirely based on a single technique, which has relatively low spatial resolution. (In functional

MRI, a signal at each location pools together processing related to a very large number of neurons; a typical location, called a "voxel," can easily contain millions of neurons.) The approach also doesn't account for the confirmation bias present in the literature. For example, researchers often associate amygdala activation with emotion and are thus more likely to publish results reflecting this association, a tendency that will increase the association between the amygdala and the domain "emotion" (not to mention that investigators might mean different things when they say "emotion"). Finally, the study makes the assumption that the 20-dimensional space of mental tasks is a reasonable decomposition. Many other breakdowns are possible, of course, and it might be even more informative to consider a collection of them at the same time (this would be like describing coffee in terms of a given set of attributes but then using separate groups of attributes).

How Should We Think about Brain Areas?

Neuroscience has agonized over this question since its modern beginnings in the last few decades of the nineteenth century, as was forcefully summarized by Paul Morgane when describing the

> ... difficulty inherent in visualizing, or even conceptualizing, organizations of neurons distributed widely throughout cortical and subcortical structures, and somehow integrated into a functional unit, without recourse to an anatomically separate integrating center or system. . . . (Morgane 1979, 14)

In other words, if not constrained to a spatially delimited area, then what?

Whereas science has proved exceedingly apt at describing modular systems, such as many found in physics and engineering, it has not made as much progress when the object of study is not as clearly decomposable. This is the case in both brain science and genetics. That both are within the realm of biology is not surprising, as biology has properties that are fairly unique and distinguish it from the physical sciences.[9]

The ideas charted in this chapter have implications to how we'll *describe* brain regions in the remainder of the book. At first, a few of their important functions will be highlighted and the brain regions will be discussed mostly on their own; this is almost inevitable didactically. At times, therefore, it will appear as if the regions themselves are responsible for the functions or behaviors described. Instead, the reader should bear in mind that brain regions participate in specific computations, functions, processes, or behaviors only when embedded within larger circuits comprised of multiple brain areas. Therefore, our understanding of the role of a specific region needs to be gradually bootstrapped so that eventually we will have a better appreciation for its functional contributions and repertoire, as we gain insight into how it interacts with other regions.

5 Emotion and Motivation: The Subcortical Players

In this chapter, we'll learn about some brain regions believed to be central to processing emotions: the hypothalamus and the amygdala. For much of the twentieth century, the hypothalamus was considered the epicenter of emotion in the brain, a position taken over by the amygdala more recently. We'll also learn about the striatum and its closely associated partners in the midbrain. The latter produce the neurotransmitter dopamine, which plays an essential role in reward and motivation. But remember, the focus is on areas to help the learning process and get us started— all mental functions rely on distributed circuits.

Scientific conferences today can be large events that draw tens of thousands of researchers. The annual conference of the Society for Neuroscience, the largest in the world in this field, gathers every fall in the United States and now draws more than 40,000 participants. Some of the most sought-after scientists are interviewed by local news stations, and it is quite a happening. But in the late nineteenth century, research gatherings were much bigger affairs (Finger 1994, 54). The 1881 International Medical Congress in London, with more than 120,000 participants, included invitations to all of Europe's royalty. One of the presentations at the meeting was by Friedrich Goltz, a professor of physiology at the University of Strasburg. Like several of his contemporaries, Goltz was interested in the localization of function in the brain. He not only published several influential papers on the problem but attracted widespread attention by exhibiting dogs with brain lesions at meetings throughout Europe. His presentations were quite a spectacle. He would take the lectern and bring a dog with him to demonstrate an impaired or spared behavior that he wanted to discuss. Or, he would open his suitcase and produce the skull of a dog with the remnants of its brain. In some cases, a separate panel of internationally acclaimed scientists would even evaluate the lesion and report their assessment to the scientific community.

In some of his studies, Goltz would remove the entire cortical surface of a dog's brain and let the animal recover. The now decorticated animal would survive, though it would generally not initiate action and remain still. Goltz showed that animals with an excised cortex still exhibited uncontrolled "rage" reactions, leading to the conclusion that the territory is not necessary for the production of emotional expressions. But if the cortex wasn't needed, the implication was that other parts of the brain were involved. That emotion was an affair "below the cortex" was entirely consistent with nineteenth-century thinking.

Victorian England and the Beast Within

In the conclusion of *The Descent of Man*, Charles Darwin wrote in 1871 that "the indelible stamp of his lowly origin" could still be discerned in the human mind, with the implied consequence that it was necessary to suppress the "beast within"—at least at times. This notion was hardly original, of course, and in the Western world can be traced back to at least ancient Greece. At Darwin's time, with emotion being considered primitive and reason the more advanced faculty, "true intelligence" was viewed as residing in cortical areas, most notably in the frontal lobe, while emotion was viewed as residing in the basement, the lowly brainstem.

The decades following the publication of Darwin's *Origin of Species* (in 1859) were a time of much theorizing not only in biology but in the social sciences, too. Herbert Spencer and others applied key concepts of biological evolutionary theory to social issues, including culture and ethics. *Hierarchy* was at the core of this way of thinking. For the survival of evolved societies, it was necessary to legitimize a hierarchical governing structure, as well as a sense of self-control at the level of the individual—it was argued.[1] These ideas, in turn, had a deep impact on neurology, the medical specialization that characterizes the consequences of brain damage on survival and behavior. John Hughlings Jackson, to this day the most influential English neurologist, embraced a hierarchical view of brain organization rooted in a logic of evolution as a process of the gradual accrual of more complex structures atop more primitive ones. What's more, "higher" centers in the cortex bear down on "lower" centers underneath, and any release from this control could make even the most civilized human act more like his primitive ancestors. This stratified scheme was also enshrined in Sigmund Freud's framework of the *id* (the lower level) and the *superego* (the higher level). (Freud also speculated that the *ego* played an in-between role between the other two.) Interestingly, Freud was initially trained as a clinical neurologist and was a great admirer of Jackson's work.

Against this backdrop, it is not surprising that brain scientists would search for the neural basis of emotion in territories below the cortex while viewing "rational thinking" as the province of the cerebral cortex, especially the frontal lobe. Let's describe now some of the brain regions historically implicated in emotion, starting with a mass of cells underneath the thalamus—the hypothalamus.

Hypothalamus and the Rage That Was Not

Researchers identified the hypothalamus (see figure 2.2) as a clear anatomical entity only in the first years of the twentieth century. Subsequently, its contributions to emotional process would be the subject of intense investigation and debate.

In the last few decades of the nineteenth century, there was a vigorous push to study how the brain interfaces with the external world, identifying areas concerned with visual inputs or the production of movements, for example. A smaller group of scientists attacked a different goal: to discover how the nervous system processes the internal world, including the control of respiration and circulation and the vegetative (also called autonomic) functions—nutritional, metabolic, and endocrine functions—required for the maintenance of life. The push to understand the vegetative functions of the central nervous system was all the more timely given the progress in describing the *autonomic peripheral* nervous system, with its dual organization containing *sympathetic* and *parasympathetic* divisions.

The peripheral nervous system contains the parts of the nervous system other than the brain and spinal cord. The autonomic nervous system, in particular, consists of the neurons that innervate the internal organs, the bloods vessels, and the glands. Its sympathetic subdivision tends to be most active during a crisis, sometimes indicated by "fight, flight, fright, and sex" (the "four Fs" memorized by American medical students). The parasympathetic division facilitates digestion, growth, immune responses, and energy storage. In most cases, the activity of the two divisions is reciprocally related; when one is up, the other is down. The sympathetic division frenziedly mobilizes the body for a short-term emergency at the expense of keeping it healthy over the long term, which is the province of the parasympathetic component. By and large, both cannot be stimulated robustly at the same time; their general roles are inconsistent.

By electrically stimulating the hypothalamus in the cat in a series of studies from 1909 to 1928, Johann Paul Karplus and Alois Kreidl demonstrated that it is involved in autonomic functions: tear secretion, salivation, sweat discharge on the footpads, rise in blood pressure, and bladder contraction (Pribram 1960; Wang 1965). But it was Walter Cannon and his students who would bring the hypothalamus to the forefront of emotion research.[2] Using early versions of an X-ray to study digestion, Cannon, then a medical student (in 1900), noticed that peristaltic contractions (which move food through the esophagus, stomach, and intestines) promptly stopped when a cat became agitated. Careful experimentalist that he was, he tested this repeatedly in the cat and established that about 30 seconds after calming, the digestive movements would start again. To Cannon, the connection between an emotional state and digestion was a clue that the nervous system played a direct role in controlling the digestive system.

Much of Cannon's research in the first decades of the twentieth century tried to uncover the workings of the autonomic nervous system. He noticed that "emotional excitement" increased diffuse activity of the sympathetic system, as if preparing the animal to fight or to escape from a predator. In his view, the responses mobilized bodily resources with the aim of preserving life under challenging and stressful conditions. If these ideas sound familiar, it's because they have been popularized by the expression "fight or flight" that Cannon himself coined.

Cannon and his collaborators performed a series of experiments to determine the role of the brainstem in emotion. They found that cats whose cerebral cortices were disconnected from the brainstem showed rage responses when coming out of anesthesia (Cannon and Britton 1925). The "rage" often was constituted of a mix of hissing, growling, fur standing on end, and other accompanying behaviors, such as pawing movements. But although these responses were fairly coordinated, they never amounted to an effective defense. Because they did not seem to reflect real anger and were not regularly directed toward the triggering stimulus, the reactions were called "sham rage."

When Philip Bard, a graduate student in Cannon's laboratory, needed a project for his PhD work, Cannon suggested that he investigate how the central nervous system engages the autonomic system.[3] Bard, who had complete independence from Cannon to steer his research, decided to try to figure out the brain level responsible for sham rage. Was it some site in the brainstem, as most believed at the time, or possibly subcortical sites in the forebrain? Bard first decorticated his cats, which produced sham rage as known. He then made various lesions to identify the critical subcortical site and saw that the reaction was unaffected until he lesioned the (posterior) hypothalamus. Both Bard and Cannon interpreted the finding as uncovering the critical site for sham rage—the hypothalamus.

The ramifications of the series of studies by Bard were far-reaching, and elevated the hypothalamus to the centerpiece of the circuitry responsible for emotion. So much so that, in the subsequent decades, the "emotional brain" was heavily anchored on the hypothalamus: if a brain region was connected with the hypothalamus, it was a strong candidate to play a notable part in emotion.

Cannon and Bard had called the emotional behaviors observed in decorticated animals "sham rage." Why "sham"? At a deeper level, emotion was thought to involve two key components: a way of acting and a way of experiencing (Bard 1934). Whereas characterizing actions is fairly straightforward (careful observation is all that is needed), subjective experience is a concept that is light-years more elusive, referring in this case to how emotion "feels." It was believed that once the cortex was removed, consciousness, including the subjective aspects of rage, would be altered and likely absent—one might be able to act, but it would be action without "anyone in there," like an automaton.[4] Although many researchers interpreted the findings by Bard, Cannon, and others to indicate that emotion was centrally dependent on the hypothalamus, Bard himself was careful to distinguish between "full-blown emotion" and what he termed a "quasi-emotion" in the case of the decorticated animal. To him, without apparent feeling (the subjective component), the observed state in an important sense should not be labeled an emotion.

How emotion "feels" was stated surreptitiously above. Yet, how the world feels to us—the multichrome colors of the sunset, the zesty taste of an unripe lime, the longing triggered by a song—is considered one of the most challenging aspects of neuroscientists' attempts to explain the brain basis of consciousness. The philosophically oriented literature spans the entire spectrum, from "impossible to explain even *in principle*" to "there's nothing inherently different" in explaining this type of mental act compared to other research questions.

The Hypothalamus Is No "Master Controller"

Since the time of Cannon and Bard, knowledge about the hypothalamus has grown by leaps and bounds, revealing that this area in incredibly multifaceted. It participates in complex homeostatic mechanisms and contributes to neuroendocrine outputs affecting brain and body glands. And it contributes to wide-ranging processes: circadian rhythms, wakefulness and sleep, stress responses, temperature regulation, food intake, thirst, sexual behaviors, and defensive behaviors. This is a staggering list of critical functions for such a small structure, as Harvey Cushing, the father of neurosurgery, rightly intuited (chapter 3); recall that he emphasized that the hypothalamus is small enough to be covered by a thumbnail. In all these processes, the region works in concert with a multitude of other sites, several of which are located in the brainstem and spinal cord, a theme we will return to later.

Textbooks picture the hypothalamus as the "head ganglion of the autonomic nervous system" (a ganglion is a mass of cells). This rubric encapsulates a hierarchical theoretical view that entails "descending" control: the area *governs* structures along the extent of the brainstem. Indeed, the hypothalamus has robust projections contacting multiple brainstem sites, including the periaqueductal gray or PAG (chapter 3), and even parts of the spinal cord; some of these areas have rather direct somatic and visceral effects on the body (figure 5.1). (Somatic refers here to the skin and skeletal muscles; visceral refers to the internal organs.) This arrangement suggests to some that the hypothalamus acts as a sort of master controller telling brainstem and spinal cord sites what to do, in line with a class-based view of brain function where, in upright humans, superior sites along the neuroaxis control inferior ones.

But the notion of a "controller" region reflects antiquated notions. No area is simply an outflow region (and thus a "head"); all areas receive inputs, too. In the present case, brainstem sites that receive projections

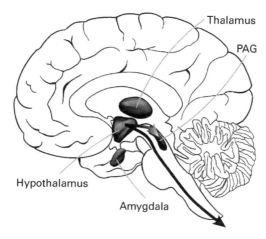

Figure 5.1
Outgoing connections of the hypothalamus. Descending projections to the periaqueductal gray (PAG) and along the brainstem are indicated by the arrow. Other structures discussed in the book include the thalamus and amygdala, shown for reference.

from the hypothalamus project back to it, illustrating the general tendency
of connections to be bidirectional in the brain (though unidirectional ones
certainly exist). Strikingly, the hypothalamus is also bidirectionally connected with large sectors of the cortex (figure 5.2)[5]—from the hypothalamus to cortex and vice versa—a property neglected in many textbooks and
forgotten by neuroscientists. So, signals from the hypothalamus go nearly
everywhere along the cortex, while the region itself listens to what's happening in the cortex and the brainstem. (In chapter 3, we discussed the
hypothalamus as part of a "mini-brain," providing contextual input to the
superior colliculus during behavioral decisions.) Why should we neglect all
of this knowledge and emphasize descending control only? More generally,
instead of outflow or inflow, it's best to characterize areas in terms of *integration* and *distribution* of signals: the more it has incoming pathways, the
more it can integrate signals; the more it has outgoing pathways, the more
it can distribute them (figure 5.3).

As mentioned, the hypothalamus participates in a staggering array of
processes. Let's briefly discuss a few of them. The region is implicated
in autonomic responses and defensive behaviors, as described by Bard,
Cannon, and their contemporaries. In the cat, for example, autonomic
responses include pupil dilation, piloerection (raised hair), accelerated

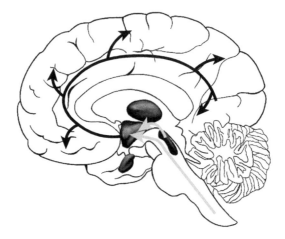

Figure 5.2
Return pathways to the hypothalamus (light gray), as well as pathways from this area
to nearly all of the cortex.

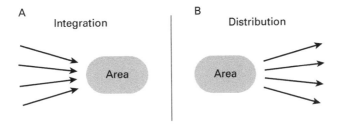

Figure 5.3
An area's potential for integration (a) and distribution (b) of signals depends on the
number of incoming and outgoing connections, respectively.

heart rate, and elevated blood pressure; defensive behaviors include hissing, growling, the ears retracting, and the animal striking with the forepaw.
The hypothalamus is also important for the "stress response," which relies
on the intricate orchestration of the hypothalamic-pituitary-adrenal (HPA)
circuit involving both brain and body (figure 5.4). Although it is located in
the brain, the pituitary is a gland that produces hormones that influence
bodily processes, including growth, blood pressure, sex, metabolism, and
pain. Interestingly, the hypothalamus is physically connected to the anterior part of the pituitary (its more glandular part) through a small tubelike
structure. The remaining component of the circuit is the adrenal glands,
which sit atop the kidneys.

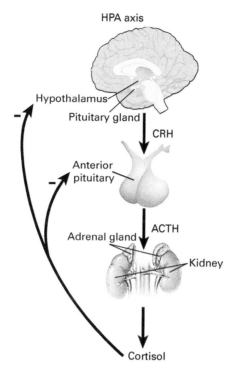

Figure 5.4
The "stress response" involving the hypothalamic-pituitary-adrenal (HPA) circuit engages a brain-body loop involving both neuronal and hormonal components. ACTH, adrenocorticotropic hormone; CRH, corticotropin-releasing hormone.

Neurons in parts of the hypothalamus actually synthesize hormones, including corticotropin-releasing hormone (CRH). The same neurons integrate stress-relevant signals they receive from multiple brain regions and launch the neuroendocrine response to stress. CRH then stimulates the release of another hormone from the pituitary, adrenocorticotropic hormone (ACTH), which in turn drives the release of glucocorticoids from the adrenal glands. The latter "stress hormones," such as corticosterone and cortisol, have a wide range of effects, including increasing blood pressure and glucose levels, as well as suppressing inflammatory and immune responses.

Both physical (for example, injury) and emotional (for example, being threatened) stressors engage a cascade of homeostatic mechanisms, of which the HPA circuit is an important part. In this manner, "stress" leads to brain and body changes aimed at redirecting energy to the central nervous system, muscle, and affected body parts. The reaction promotes *homeostasis*—that is, the maintenance of the state of the body within an adequate range of functioning—somewhat like an automatic pilot system maintaining all the variables linked to a plane within a proper range.

The overall stress response is intended to handle the insult but be of limited duration, thus minimizing sequelae to the body; enhanced cardiovascular, endocrine, immune, and visceral activity takes a toll on the body's physiology, impacting health.[6] Although stress responses are often triggered transiently, the system can also get stuck in an "on state." Intriguingly, prolonged depression seems to be accompanied by an exaggerated and prolonged reaction that is like a "stress response." In fact, depressed patients exhibit behavioral patterns that are reminiscent of those observed in rats administered with CRH, the hormone produced by the hypothalamus (figure 5.4). More generally, the impact of stress on health is difficult to exaggerate. Traditional cardiovascular risk factors include smoking and obesity, of course. It may come as a surprise that stress-related factors are associated with comparable or even higher peril. For example, high work stress quadruples one's chances of developing cardiovascular disease, problematic marriages multiply by three the risk for heart problems, and caring for a partner with Alzheimer's disease doubles cardiac hazard.

Gasping for Air and the Amygdala

Immediately following the inhalation of CO_2, patient SM began breathing at a rapid pace and gasping for air.[7] Approximately eight seconds following the inhalation, her right hand started waving frantically near the air mask. At 14 seconds, SM exclaimed, "Help me!" while her right hand gestured toward the mask. As this was happening, her body became rigid, her toes curled, and her fingers on both hands were flexed toward the ceiling. The experimenter immediately removed the mask from SM's face. As soon as the mask was removed, SM grabbed the experimenter's hand and in a relieved tone said, "Thank you." The skin on her face was flushed, her nostrils were flared, her eyes were opened wide, and her upper eyelids were raised. Fifteen seconds later, SM's breathing began to return to normal; she let go of the experimenter's hand and then said, "I'm alright."

This was the first time that patient SM experienced fear in her life, even though she was in her mid-forties. Yes, you read it correctly—the first time.

SM had grown accustomed to being tested. For over a decade she had volunteered in a plethora of experiments by researchers studying the amygdala. What makes her so special to them is that she's one of probably a handful of persons in the world that has a nearly complete natural lesion of the amygdala in both brain hemispheres (the damage results from a process of natural calcification of the tissue).

Undoubtedly, the amygdala is one of the stars of the brain as far as media coverage is concerned. It seems that no more than a few months pass without a major media outlet discussing it in an article: "The Amygdala Made Me Do It"; "The Secret to a Good Scream"; "Humans, Like Animals, Are Fearless without Amygdala"; "The Political Brain"; "Fear and Anger Heard Deep inside the Brain"; "How, but Not Why, the Brain Distinguishes Race." Titles like these are par for the course (these were found searching for the keyword "amygdala" in the *New York Times*).

Back in 1819, the German anatomist Karl Burdach described a mass of gray matter seen in slices through the temporal lobe (figure 5.5). He called the almond-shaped structure the amygdala, for the Greek word given to the nut. Electrical stimulation studies starting in the early 1950s revealed that the area is involved in autonomic responses, including cardiovascular, respiratory, pupillary, and bladder responses (Kaada 1960, 1972). Its role

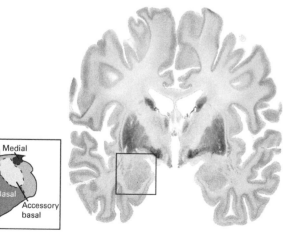

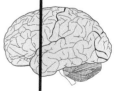

Figure 5.5
The human amygdala. In this coronal slice, the amygdala of the left hemisphere (indicated by the rectangle) is expanded on the left to schematically reveal its subnuclei. The brain on the right shows the approximate level of the high-resolution slice.

in "aggression" and "fear" was also noted. It was the link to the latter that rocketed this region to stardom. In 1979, investigators reported that the amygdala is important for learning the *affective significance* of a stimulus (Kapp et al. 1979), and studies in the subsequent decades firmly established its necessity during this process.

The amygdala is critical for learning the aversive significance of items that, at first, are neutral. The process can be studied by employing techniques of classical conditioning, also called Pavlovian conditioning (after Ivan Pavlov's contributions to the understanding of these learning mechanisms), or fear conditioning. In aversive classical conditioning, the subject is exposed to a neutral conditioned stimulus (CS), such as a tone or a light, which typically co-terminates with an aversive unconditioned stimulus (UCS), such as a foot-shock, which is unconditioned because no learning is needed to establish its aversiveness. With training, the CS (tone/light) acquires aversive properties and, when subsequently presented alone, elicits a "conditioned response." In rodents, this response involves freezing behavior (the animal remains still), alterations in autonomic nervous system activity, release of stress hormones (as discussed in the context of the stress response), analgesia, and facilitation of reflexes (they startle rather easily). Subsequently, conditioned responses can be suppressed, or at least largely reduced, if the conditioned stimulus is repeatedly presented alone, a phenomenon called "extinction." In other words, the animal learns that the CS is not aversive anymore.

Like most subcortical regions, the amygdala is highly heterogeneous (figure 5.5), even more so because it contains an entire sector that is "cortex like," not in the sense that it contains multiple cell layers but because of its connectivity pattern. Pathways from the lateral and more inferior sector, called the *basolateral amygdala*, reach almost all of the cortex, and most of the connections are bidirectional (figure 5.6a). Another component that we'll discuss here is more centrally located and simply called the *central amygdala*, which has a qualitatively different set of output pathways (figure 5.6b). As can be intuited, these two amygdala sectors are functionally quite distinct.

The basolateral amygdala is a portal for sensory stimuli, receiving signals related to auditory, visual, and somatosensory information, allowing both conditioned and unconditioned stimuli to be registered. Indeed, the convergence of both types of stimuli enables this sector to form associations between the two—for example, between a light and a foot-shock. This

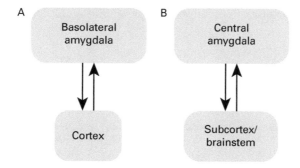

Figure 5.6

Patterns of amygdala anatomical connectivity. (a) The basolateral amygdala is strongly interconnected with most of the cortex, although some of the connections are relatively weak (such as those with the lateral prefrontal cortex). The basolateral amygdala is also interconnected with subcortical areas (not shown). (b) The central amygdala has a very different pattern of connectivity, one that is heavily interconnected with the subcortex and brainstem. The basolateral amygdala also receives some projections from the cortex, such as from the cingulate cortex (not shown).

basolateral amygdala sector is the part that is critical for aversive *learning*. How do we know this? First, lesion or general inactivation of this subarea prevents associations from being established. However, with these kinds of studies it is hard to map the critical location because tissue removal often impacts adjacent gray matter in ways that are difficult to quantify. Second, cell responses elicited by the CS in the basolateral amygdala are modified after pairing with the UCS. Cells that initially respond little to the CS increase their firing with conditioning, which is interpreted as a neuronal correlate of learning the CS-UCS pairing (figure 5.7). Third, recent studies employing modern genetic techniques confirm that plasticity in the basolateral amygdala is *necessary* for aversive learning. Finally, aversive conditioning can be prevented when sensory inputs to the lateral amygdala are precisely blocked.

Let's consider the central amygdala now. This sector plays a key role in the generation of conditioned responses—referred to as "emotional responses"—that affect the body. How does the central amygdala accomplish this? Its impact on the body depends on anatomical projections to regions that have a more direct effect on autonomic and skeletomotor responses (figure 5.6b). For example, the central amygdala projects to the PAG, which, as we saw in chapter 3, contributes to animal's defensive behaviors—both passive (such as freezing in place) and active (such as moving away) behaviors. It also projects

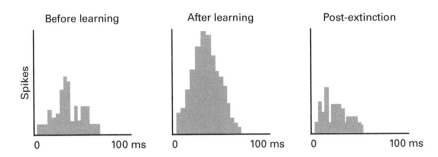

Figure 5.7
Aversive conditioning and amygdala cell responses before learning, after learning, and post-extinction. Following conditioning, responses elicited by the conditioned stimulus (such as a tone) become stronger (shown by the spikes). After the relationship between the tone and shock is discontinued (a process called extinction; see chapter 10), cell responses decrease again.

to the hypothalamus, which as we saw contributes to autonomic system activation. And projections to other parts of the brainstem potentiate motor responses leading to the so-called startle reflex.

The central amygdala also contributes to conditioned responses by changing the availability and distribution of neurotransmitters across large swaths of the brain, which is accomplished via projections to brainstem nuclei that are neurotransmitter factories (chapters 2 and 3). Although small, these chemical plants project rather broadly and diffusively across the brain, thus having a sizable mark on ongoing processing. Keep in mind that neurons throughout the brain aren't sensitive to all neurotransmitters equally; instead, they show preference for a few of them depending on the receptors that they contain (receptors are sites where the neurotransmitters can bind and thus affect the communication between neurons). Therefore, the chemical release of neurotransmitters across the brain can have a more selective impact on neuronal function than the diffuse projection of fibers would suggest.

One of the targets of the central amygdala in the brainstem provides a good example. The area is called the *locus coeruleus* (literally "blue spot," given its color as seen in fresh human tissue from the presence of melanin pigment), and it synthesizes the neurotransmitter norepinephrine. The locus coeruleus, which can also be engaged by other regions besides the amygdala, is rather small, containing less than 50,000 neurons (a tiny fraction of the more than 85 billion neurons in the brain). Despite its small size, it packs a big punch as its fibers project widely throughout both the subcortex and the cortex. Norepinephrine, like other neurotransmitters, is important as a *modulatory* substance, which is to say that rather than producing direct excitatory or inhibitory effects, it nudges up or down the effects produced by other cellular communication (itself dependent on neurotransmitters). Salient and arousing stimuli, like loud noises, are very effective at eliciting a burst of activation in the locus coeruleus, leading to the release of norepinephrine in the regions' many cortical and subcortical targets. Researchers have thus proposed that the locus coeruleus functions as the brain's analog of the adrenal gland, augmenting the processing of *motivationally* relevant stimuli and preparing the brain to handle the potential insult.

Remarkably, the central amygdala targets not only the locus coeruleus but several other brainstem sites producing the neurotransmitters dopamine, serotonin, and acetylcholine.[8] Accordingly, conditioned stimuli that engage the central amygdala, by their effect on multiple brainstem sites, have diverse effects on signal processing throughout the brain. By being paired with an unconditioned stimulus and acquiring affective significance, a once-neutral stimulus gains the ability to mobilize the body and the brain to, hopefully, adequately handle the situation at hand (figure 5.8). At the same time, it will come as no surprise that that dysfunctions of this system are believed to contribute to mental disorders, such as anxiety.

Back to patient SM, who experienced fear for the first time in her forties. Before the CO_2 inhalation experiment, it was believed that patients like her experience little or no fear. In fact, cases like SM's, and the large research

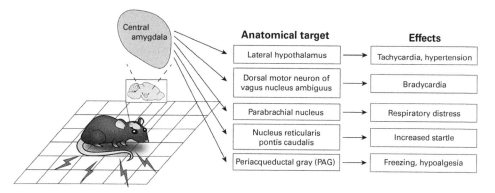

Figure 5.8
Central amygdala targets and some of the associated physiological changes.
Source: Pessoa (2018a) and inspired by Davis (1992).

literature on the involvement of the amygdala in aversive classical conditioning, have led to a fairly entrenched view of the amygdala as a hub for
fear. So, how did SM experience fear when breathing carbon dioxide? There
are many different types of fear, not just one. Carbon dioxide is unique in
that chemoreceptors sensitive to its presence engage sensory pathways that
project to the brainstem. That is to say, CO_2-related fear depends on extra-
amygdalar circuits. Currently, the most likely scenario implicates the PAG
in the upper brainstem. (That excessive CO_2 should provoke fear is perhaps
not so surprising. After all, increasing CO_2 indicates decreased oxygen levels
from pulmonary malfunction or reduced ambient oxygen.)

More broadly, the amygdala, important as it is for some fear-related
learning, is not the sole region that is *critical* for it.[9] Indeed, it is increasingly
appreciated that fear conditioning engages a broader and more complex circuit than initially thought. For example, sites such as the thalamus and the
cortex undergo changes (also called plasticity) during aversive conditioning
and contribute to enhancing responses to conditioned stimuli. Without the
participation of these regions—say, if they are experimentally inactivated—
the learning process is compromised.

The Automaticity of It All

Until the 1990s, to study the physiology of the amygdala required invasive
recordings performed in animals. The development of functional magnetic
resonance imaging (fMRI) was revolutionary, allowing brain activity to be
recorded safely across the entire human brain—as long as participants could
stay in a constrained and noisy tube without moving their heads for 30 to
60 minutes, if not longer! The availability of MRI machines for research purposes and not just clinical usage opened the door for investigators to probe
all sorts of mental phenomena, from basic perception to moral judgments.

Against this backdrop, researchers were eager to investigate the amygdala. Although the MRI scanner is a very constrained environment, it is easy
to display images to participants and to collect simple responses through
button-pressing. Neuroscientists reasoned that when people see images
with emotional content, responses in the amygdala would ensue. Images
of mutilation, accidents, and disgust-inducing vomit or feces were quickly
adopted as ways of studying "emotion perception." It was found that the
amygdala, along with several other regions, produced stronger responses

to emotion-laden compared to neutral images. Investigators also studied responses when subjects viewed pictures of facial expressions, including expressions of fear, anger, surprise, disgust, and happiness. Although these images are not overtly emotion-inducing, investigators found that responses to fearful faces were stronger than neutral ones, for instance. Once again, researchers were quick to point out that the amygdala is "tuned" to fear. (The amygdala responds vigorously to all kinds of faces, including those with a happy expression, as well as neutral ones.)

Remarkably, stronger responses to fearful faces were detected even when they were flashed for a split second (presented for around 1/30th of a second). Now, if you briefly flash a fearful face and then follow it with a picture of a neutral face, participants actually seem to miss the initial one and report only seeing the neutral face. In such cases, the second stimulus is said to "mask" the first. What would happen in the brain in such cases? Would it respond in a manner reflecting that the participant was physically stimulated first with a fearful face? Put another way, would the brain register the stimulus even when the person was not clearly aware that the first face was shown? This would be quite wonderful to psychologists interested in all sorts of "priming effects." (Priming effects come in all sorts of shapes and sizes. For example, when primed with the word "nurse," a participant will react a little faster to a semantically related word, such as "doctor.") Quite a stir was generated when researchers measured responses in the amygdala even when participants were not aware of them.

Studies like this led to the view that amygdala responses are *automatic*: generated whether or not a person pays attention to a stimulus and even if one is unaware of it. To determine the role of attention, investigators capitalized on procedures that vision scientists had concocted to manipulate how attention is allocated and "used up." For example, a participant would be asked to gaze at a central cross while a sequence of letters is presented to the left and another to the right. If the task is to consider only the left stream and to indicate if the letter "X" appears there, naturally, people focus their attention as much as possible on this stream. Researchers call the stream on the left "attended" and the one on the right "unattended." Neuroscientists found that faces showing the expression of fear were processed more strongly even when they were not explicitly attended. (Note that it is possible to move one's focus of attention without explicitly moving one's eyes. The dissociation between attention and gaze direction, which one may use in not giving away that they are monitoring someone, was described by the nineteenth-century physicist and physiologist Hermann von Helmholtz. He was not only interested in the foundations of thermodynamics but equally so in the mathematics of the eye and theories of vision.)

Masking and attention-manipulation results led to what a colleague of mine, Ralph Adolphs, and I dubbed the "standard hypothesis" (Pessoa and Adolphs 2010): Emotional stimuli are processed initially by a dedicated, modular system centered on the amygdala that operates rapidly, *automatically*, and largely independently of conscious awareness. What's more, defects in this system are suggested to underlie phobias, mood disorders, and posttraumatic stress disorder. Overall, it was a very influential theoretical framework informing both basic research and applied studies focusing on mental health.

In the early 2000s, my own neuroimaging work sought to understand what exactly was meant by "automaticity." In the field of psychology, the use of the term often referred to processes considered effortless, nonconscious, involuntary, or possibly obligatory. This general definition allowed diverse phenomena involving different processes, such as detecting very salient visual stimuli, well-practiced cognitive or perceptual-motor skills, and even some forms of social information processing ("Is this person white or Black?"), to be viewed under a single theoretical umbrella. But demonstrating automaticity proved to be quite slippery. A good example is the intuitive idea that "visual onsets" (like a letter appearing in a previously blank location) capture attention automatically. Although initial studies suggested that abrupt and salient visual stimuli are involuntarily registered, subsequent experiments revealed that they aren't, illustrating a common pattern of initially reporting a phenomenon to be automatic and later, on more refined experimental probing, discovering otherwise. Why? If an experiment effectively "uses up" attention by making one condition very challenging, one's ability to process other things (even an abrupt visual onset) is pretty much eliminated—much like a driver will miss a crossing pedestrian right in their line of sight if consumed by their phone.

We discovered the same with the perception of emotion-laden stimuli. In a series of studies in my laboratory, then at Brown University, we found that when attention is really pushed by a challenging task that consumes it, unattended stimuli do not produce detectable differences based on emotional content (like a fearful face generating a stronger response). A series of other laboratories across the world found similar results. So, while emotion-laden stimuli are clearly more potent than neutral ones, they are not so strong as to be processed "no matter what."

Back to visual awareness. Researchers had agonized with various ways to study memory for a long time and wondered how the questions used to interrogate participants affected their response. For example, someone may say whether or not they "remember" a previously shown card depending on how certain they feel about it and on their take on the experimenter's intentions. Perhaps experimenters want us to say "yes" only if we are very certain, or perhaps it's okay if one experiences a vague feeling of familiarity. Faced with this response criterion problem (at what point does a participant say "yes"?), investigators sought methods that could measure memory ability while taking this factor into account. The same applies to visual awareness studies. In the masking procedure described previously, participants were asked if they ever saw a fearful face, for example, after coming out of the scanner. But how should participants gauge their yes-versus-no point? Having performed many masking studies myself, I can say that when two stimuli are shown back to back, there's something a bit jarring going on (even if an initial neutral face is masked by another neutral face). When we performed functional MRI studies in my lab using the masking paradigm, we adopted the same approach used in memory research that accounts for a participant's inherent tendency to say or not to say "yes" (their response criterion). With this methodology, we didn't find evidence of unconscious processing of emotion-laden stimuli.[10] In the end, based on a comprehensive review of the literature, Ralph Adolphs and I proposed that, while appealing, the "standard hypothesis" isn't tenable.

Although not automatic, visual processing of emotion-laden stimuli is quite remarkable. Areas in the occipital and temporal cortex that process visual attributes are strongly engaged by emotion-laden stimuli. That is to say, when the brain processes visual content with emotional significance, the visual cortex responds more vigorously. It is as if the "volume" of the stimulus is turned up when it is emotional, with a lot of visual cortex reflecting this. In one of the experiments from my lab, people in an MRI scanner watched short clips containing rapid, flashed-up sequences of images (Lim, Padmala, and Pessoa 2009). Among them were human faces and, after them, either an image of a house or a skyscraper. Participants had to determine which of these two scenes was present in the clip—a task that was very difficult because they had to pay attention to both the faces and the subsequently presented building. But our experiment had a twist. Before viewing the clips, half of the participants received a mild electric shock while viewing a series of skyscrapers, but they never received a shock when viewing houses; conversely, the other half watched a series of houses appear, paired with the same mild shock, but never experienced a shock with skyscrapers. This is a version of classical conditioning, of course, and links an initially neutral stimulus (a nondescript picture) with the emotional meaning of the unpleasant stimulus (the shock). The outcome: Participants conditioned to the skyscrapers detected them better than houses; conversely, participants conditioned to houses detected them better than skyscrapers. And in each case, responses in the visual cortex were stronger for the type of stimulus (house or skyscraper) to which participants had been conditioned. This study illustrates how perception is not passive at all. Rather, it involves picking up on the significance of objects and determining how they are processed. Vision is never neutral—it is always pregnant with meaning.[11]

Fear Myopia

In the past few decades, so much of the research on the amygdala has been about fear, fear, and fear that researchers are increasingly calling it a form of tunnel vision to a large extent generated by the enormous success of the study of classical fear conditioning. Yet the fear-centric view of the amygdala is in need of an update. The same neurons and circuits that participate in aversive learning also respond to rewards and support conditioning with positive items, including food- and sex-related stimuli. More generally still, as the region participates in a host of functions, focusing on its contributions to signaling threat in the service of defensive behaviors is extremely myopic.[12]

Selection of information for further analysis is a key problem that needs to be solved for effective learning and arguably many other behaviors. How can a limited-capacity information processing system that receives a constant stream of diverse inputs—such as the nervous system—be designed to selectively process those inputs that are most significant to the objectives of the system? The amygdala seems to be intimately involved in solving this problem. Put another way, the region cares about *selective information processing*. According to this broader perspective, the amygdala is not an "emotion structure." Whereas some of its functions can be described using this optic, its contributions extend far beyond it and include attentional functions and even decision making (see chapter 7).[13]

Take a study in which rhesus monkeys chose between "saving" a liquid reward and "spending" the already accumulated reward immediately. On a given trial, the monkeys could either save the reward for the future, in which case it would accrue some interest, or consume the accumulated pot. Behaviorally, the monkeys kept track of the accumulated rewards over successive save-trials and based their choices on this information. Remarkably, cell activity of neurons in the basolateral amygdala predicted the behavioral actions of the monkeys—to save or to spend. This and other studies provide compelling evidence that the amygdala *codes* at least some simple "economic choices" and in fact contributes to decision making more generally. For example, in a functional MRI study in humans, the signal measured in the amygdala could be used to predict subjects' saving plans value up to two minutes before the saving goal was achieved. A growing literature is uncovering how the amygdala contributes to plans, including the formation and execution of economic saving strategies that are future oriented.

So much for the notion of the amygdala as a primitive "fear module."

Bringing in Motivation

James Olds arrived at the lab on Sunday morning to see if everything was ready for the experiment that he and his undergraduate assistant would perform the next day.[14] He placed the rat in an open field, attached the stimulation electrode, and, using a handheld button, applied electrical stimulation—trains of 60-hertz (Hz) sine waves lasting for a quarter of a second. The rat kept returning to the area of the open field where the last shock had been given, as if it had liked the stimulation. As Olds anthropomorphized later, the rat seemed to be saying, "I don't know what I just did, but whatever it was, I want to do it again." He had just discovered the brain's "reward system"—as the finding has been described by neuroscientists.

Olds had not planned to study reward, and the findings were completely serendipitous. At the time, many investigators were interested in the "activating system" in the brainstem discovered a few years previously to be critical for wakefulness. As recounted later, it was his lack of aim (his electrical stimulation landed off target) that lead to the unexpected discovery.

However, it was very much to his credit, and in no small part to the graduate student, Peter Milner, that the effects of electrical stimulation were conclusively linked to reward processes and not to potential confounds. The
publication of the study in 1954 by Olds and Milner led to an explosion of
work in the field of *motivation*, a research area that focuses on understanding how animals seek rewards. What are animals' preferences, how much
value do they assign to certain goods, and how much effort are they willing
to exert to obtain a reward?

The cleverest manipulation employed by Olds and Milner was to let the
animal "self-stimulate" (figure 5.9).[15] When placed in a box, the rat was free
to move around while the electrode was positioned to stimulate a specific
brain structure if the animal stepped on a pedal. After the very first electric
stimulation, the rat began to search eagerly. It would then sniff all corners
of the box and quickly manipulate objects in its path until it, accidentally,
stepped on the pedal a second time. After the second or third press, it would
then cease to wander; it would then step on the pedal once or twice every
second. So the animals quickly learned to self-stimulate by pressing the
lever that caused stimulation. When the electrode was placed in the hypothalamus, one of the most reliable regions leading to this behavior, rats
would self-stimulate at a rate of approximately 2,000 presses per hour; and

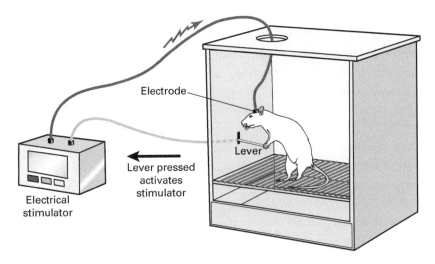

Figure 5.9
Electrical self-stimulation of certain brain sites is strongly rewarding, as inferred from
the animal's behavior.

if some parts of the midbrain were stimulated, the rats would step as many as 7,000 times an hour! Whereas the rate of lever pressing varied considerably based on the stimulation site, many locations affected the animals' behavior. In the case of an early experiment, of the 76 electrodes implanted to randomly sample locations in the midbrain and the forebrain, in 36 of them rats pressed the lever. The other brain sites were not all neutral, however. At 11 of them, the animal clearly avoided the pedal, revealing that stimulation there was aversive.

The Midbrain's Dopamine Fountain

Some items are inherently appetitive, such as particular foods. Others are initially neutral and acquire significance once associated with a positive outcome. This may occur, for example, through classical conditioning, as discussed earlier in the case of aversive items, and now in the appetitive domain. The now-significant stimulus, or conditioned stimulus, will influence future behaviors. Conditioned stimuli attract attention, which is potentially advantageous because it might draw the animal closer to sources of natural rewards, such as a favored fruit. Animals will work to obtain a conditioned stimulus; they will press levers or run around a wheel to be able to elicit the conditioned stimulus. Why is this surprising? Recall that the CS itself is not inherently rewarding—it is associated with an item that is. (For related reasons, conditioned stimuli have the adverse effect of attracting humans toward drugs of abuse, as well as legal drugs such as cigarettes. In such cases, visual or olfactory cues can become powerful conditioned stimuli.)

What are some of the brain regions that are important for motivated behaviors? A sector of the lower part of the striatum has attracted enormous research interest. This area, called the *nucleus accumbens* (or simply accumbens), has neurons that are rather sensitive to the neurotransmitter dopamine, as the dendrites of neurons there are peppered with receptors that favor the binding of dopamine (figure 5.10). And, like the amygdala, the accumbens is a favorite of the media: "Can You Get Over an Addiction?"; "This Is Your Brain on Drugs"; "Off Drugs, onto the Cupcakes"; "Are You Programmed to Enjoy Exercise?"; "How Carbs Can Trigger Food Cravings"; "Risky Rats Give Clue on Brain Circuitry behind Taking a Chance"—these are all article titles found by searching for the keyword "accumbens" in the *New York Times*.

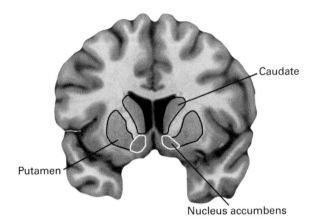

Figure 5.10
Approximate location of the nucleus accumbens in the human brain. The accumbens is located at the ventral part of the striatum; the striatum includes the caudate and the putamen.

In the early 1980s, the idea that dopamine is important for motivation and linked to reward and reinforcement took shape, spurring a tremendous amount of work to dissect how it promotes this type of processing. In the accumbens, dopamine is important when an animal encounters novel, salient, and reward-related stimuli. As seen in chapter 3, despite frequent attempts to link dopamine to reward, neurotransmitters are involved in multiple functions, and their effects vary based on the brain region or circuit they act on, as well as the behavioral context of the animal (is it hungry? under threat?). There is no such a thing as a "reward molecule"—this can't be emphasized enough.

Although accumbens neurons are strongly influenced by dopamine, this chemical is not endogenous to this region. Projections from the substantia nigra (chapter 3) release it there. The dopamine-containing neurons of the substantial nigra form a continuous band that extends into adjacent parts of the midbrain (called the ventral tegmental area), which also synthesize this molecule. Collectively, this "dopaminergic midbrain" is the origin of anatomical pathways that play key roles in motivated behaviors.

The midbrain fibers that reach the accumbens also target the superior part of the striatum (containing the caudate and putamen) through a circumscribed pathway directed at the dopamine-receptor-rich striatum. A separate bundle of fibers reaches the cortex, but not in a regionally organized manner. Instead, the projections course through the entire cortical mantle, being particularly well represented in the frontal cortex but also extending into the parietal and temporal cortex (and sparsely only into the occipital cortex). This midbrain-to-cortex white matter projection system has a far-reaching impact on brain processing.[16]

Is This Enough Reward?

The link between dopamine and the processing of reward began to be worked out in the 1980s, but the publication of a paper in the journal *Science* in 1997 by Wolfram Schultz and colleagues ushered a second gold-rush era of motivation research in neuroscience (the first being triggered by the self-stimulation study by Olds and Milner in 1954). This second wave, which is still ongoing, is guided by an elegant concept known as the *reward prediction error*: a mismatch between actual and expected rewards, and the link between this mismatch and learning.

Rewards produce learning.[17] Pavlov's dog hears a bell, sees a sausage, and salivates. If the bell-sausage pairing is repeated often enough, the dog will salivate merely upon hearing the bell. We can say that the bell predicts the sausage, and that's why the dog salivates. This type of learning occurs in a fairly passive fashion, as the dog only needs to be present for it to take place. In contrast, operant conditioning, another basic form of learning, requires the animal's participation. Thorndike's cat runs around a cage until it happens to press a latch and suddenly gets out and can eat. (Edward Thorndike, a pioneer in the study of learning, described this type of learning in 1911.) The food is great, and the cat presses again, and again. Operant learning requires the subject's own action, otherwise no reward will come and no learning will occur, just as Olds and Milner's rats needed to lever-press to receive electrical stimulation. Both Pavlovian and operant learning are related to "prediction errors," as we'll see.

What type of information is useful during learning? Imagine a comparison mechanism between a predicted and an actual reward; say, a rat anticipates its preferred food morsel and receives a much less pleasant one (figure 5.11). Logically speaking, one of three cases must be true: The reward is better than, equal to, or worse than its prediction. Here, we can think of the comparison as involving the *value* of an item, such as the preferred morsel's tastiness; the "predicted value" can be based on one's prior experience. If the reward

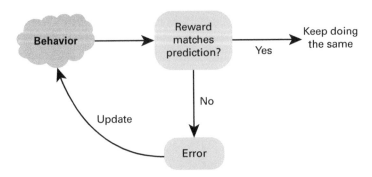

Figure 5.11
Predicting reward and learning. Only when a prediction differs from the outcome is it necessary to update expected outcomes. Otherwise, no changes in behavior are necessary.

is *different* from its prediction, we say that a prediction error occurred. How is this related to learning? If the reward is exactly as predicted, we've learned nothing new, so there isn't any need to update expectations about the future. However, if there's a mismatch, it behooves us to update our expectations. On the one hand, if the reward is better than predicted, we update our prediction to predict "higher value" in the future. Importantly, we should do more of the behavior that resulted in that reward. On the other hand, if the reward is worse than predicted, we update in the opposite direction so as to predict "lower value" next time around, and we should be less likely to repeat the behavior. We can thus think of learning from prediction errors as a type of learning from one's mistakes, as the mismatch helps to guide the learning process in the right direction. If no error occurs, the behavior stays put and is not updated; otherwise, it is updated in the appropriate direction—supposing that doing more to attain rewards is a good idea! The same logic applies to many other kinds of learning. Imagine being a singing student and you are told that you are off-key—say, half a note too high. When you repeat the exercise, you should strive to produce a lower key. Of course, if you've hit the desired note, there's no need to change anything.

What the study by Schultz and colleagues revealed was that the response of dopamine-containing neurons in the striatum appeared to signal a reward prediction error, *not* reward per se. Neurons did not fire more vigorously when a reward was obtained, but instead they responded if there was a discrepancy between the actual reward and what the animal expected, based on previous trials. Firing increased if the actual reward was greater than antici-pated, or decreased if it was smaller than anticipated. In one sense, the find-ing was surprising, as many investigators in the field thought of dopamine as signaling reward itself (something that by now the reader knows makes little sense!). But, on the other hand, the type of signal uncovered—the prediction error—was precisely what was formally predicted in mathemati-cal models of learning developed in the 1970s. For neuroscience to hit on experimental findings that could be described mathematically, albeit using fairly simple equations, was in itself quite remarkable, as such success sto-ries have been the province of physics or other more quantitative scientific research areas.

The role of dopamine and of reward prediction errors continues to be actively investigated. They have major implications for the understanding of addiction and other disorders of motivation; as alluded to above, it is prob-ably not always a good idea to seek to maximize reward. At present, there are multiple competing ideas, and they continue to be revised and refined as ever-more powerful techniques become available to neuroscientists.

Although there is so much we don't know about the brain, clearly the vol-ume of existing knowledge is far from trivial. In this chapter, we covered considerable ground as we started learning about brain regions important for emotion and motivation. In chapter 6, we'll focus on regions of the cortex. In chapter 7, we will turn to cognition. At that point, we'll finally be able to start putting things together to better understand how networks of brain regions bring about cognitive-emotional behaviors. In particular, in chapter 11, we will discuss how the amygdala works closely with other areas as part of large-scale circuits in "learning to forget" aversive memories.

6 Emotion and Motivation: The Cortex Comes to the Party

In this chapter, we'll visit several cortical brain regions that contribute to emotion: the cingulate, the insula, and the orbitofrontal cortex. Early electrical stimulation studies of the brain of patients undergoing neurosurgery revealed that passing weak electrical currents through these sites could elicit feelings and sensations pregnant with emotion, as well as alter physiological responses such as heart rate and pressure, pupil dilation, and respiration, all of which are observed during naturally occurring emotion-laden episodes. We'll see that these regions are interlinked with the amygdala, the hypothalamus, and other brainstem areas that both sense and influence the body.

The Warren Anatomical Museum in Boston is one of the last surviving anatomy and pathology museum collections in the United States. It houses the skull of Phineas Gage, who died in 1860 but whose life-defining moment took place when he was 25 years old: a tamping iron flew through his brain, but he survived to become one of the "great medical curiosities of all time" (Macmillan 2004).[1] Not only has Gage's case become a fixture of psychology and neuroscience textbooks, but he is even known in popular culture for the "personality change" that ensued after the accident. Although the details are murky, Gage is described as hardworking and responsible before the injury, but "fitful, irreverent . . . capricious and vacillating" afterward, as reported by the town doctor, John Harlow, who examined him. Harlow's proclamation that Gage was "no longer Gage" has captured the imagination of medical doctors and scientists alike.

In the nineteenth century, the strong dichotomy between subcortical ("primitive") and cortical ("advanced") brain parts relegated most of emotion to the subcortex. Yet, not entirely, as illustrated by Gage's medical case. The iron rod, all 1.1 meters of it, hit Gage from below, entering the left side of his face in an upward direction, possibly fracturing the cheekbone;

it passed behind the left eye, through the left side of the brain, and out the top of the skull through the frontal bone.[2] In the early 1990s, using Gage's skull, the damage was investigated with modern imaging techniques to estimate the likely trajectory of the projectile. Although there is dispute whether the lesion affected both sides of the brain or only the left hemisphere, there is no question that the prefrontal cortex was compromised. At the time of the accident, insofar as the behavioral changes were not interpreted to be cognitive, such as those related to language or problem solving, they were deemed to be emotional in essence, thereby strongly implicating the cortex in this type of processing. The ramifications of Gage's lesion for how the brain brings forth the mind had a substantial impact on nineteenth-century thinking.

There's another reason emotion was linked to the cortex in the nineteenth century: consciousness. As we saw in chapter 5, emotion *feels like something*—such as when one is in a rage or in a state of extreme happiness—and, in fact, many researchers consider this property one of its defining properties. Thus, emotion is frequently conceptualized as tied to one of the "highest" components of the mind: conscious awareness. And, to the extent that consciousness and emotion were interlinked, it was natural to conceive of the latter as involving the cerebral cortex, too.

In the United States, William James, the brother of the famous novelist Henry James, was one of the main exponents of psychology as an independent scientific discipline.[3] In a paper published in 1884, James proposed that emotion depends on sensory and motor centers in the cortex. For James, emotion did not depend on separate processes specially devoted to this mental faculty. Instead, it was tied to the changes that occur in the *body* during a triggering event, such as in his famous example of encountering a bear in the woods. For him, then, the feeling of the changes in the body that follow an "exciting fact," as they occur, *is the emotion*. Contemporaneously, the Danish medical researcher Carl Lange outlined a very similar idea whereby emotional events are "brought to consciousness in that they are brought to the centers of taste and vision in the cortex."

We see that early attempts to understand the neurological basis of emotion clearly encompassed the cortex. Nevertheless, these ideas were formulated only in the most general terms, which of course isn't surprising given how little was known about the brain. It would take many decades before the contributions of the cortex to emotion would begin to be elucidated.

Before delving into the brain sectors discussed in this chapter—the cingulate cortex, the insula, and the orbitofrontal cortex—I offer a reminder. As in chapter 5, the text will keep the areas/sectors largely separate from one another for expository purposes. Again, keep in mind that they work jointly. Chapters 8 to 11 will describe how to put them together in a more principled way.

Electrically Stimulating the Brain

"I was afraid and my heart started to beat," said the patient on the operating table (Vogt 2009, 12).[4] The neurosurgeons had just passed mild electrical stimulation trough the anterior part of the cingulate cortex (figure 6.1). Upon stimulation of similar locations, other patients had reported intense and overwhelming feelings of fear, including one patient who reported a sensation so intense as to be described as the feeling of imminent death. How could patients describe their experience in the middle of neurosurgery? As it happens, because the brain contains no pain sensors, the procedure is frequently performed under local anesthesia, and patient feedback is invaluable.

That the human cortex is electrically excitable was first established by Roberts Bartholow in 1874, soon after the experiments by Fritsch and Hitzig using dogs (see chapter 4).[5] Bartholow stimulated the cortex of a dying "feeble-minded" girl whose brain was ulcerated so badly that her pulsating brain could be seen (incisions had already been made to allow the pus to

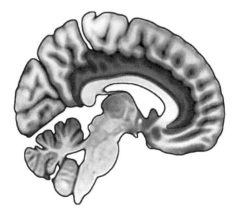

Figure 6.1
The cingulate cortex is shown in a darker shade along the medial surface of the brain.

escape). He found that, upon stimulation, she moved her limbs and felt tingling sensations, both on the opposite side of her body. Although Bartholow felt that he could introduce needles into the patient's brain "without material injury," he was severely criticized in both the United States and Europe for conducting such an experiment on a person.

Harvey Cushing, who was quoted in the context of the hypothalamus in chapter 3, was one of the pioneers of electrical stimulation during neurosurgery in the beginning of the twentieth century. Such work wasn't performed out of scientific curiosity but to aid in the planning of the operation, which was then performed to remove tumors or sites generating epileptic seizures. To the extent possible, surgeons sought to avoid removing cortical tissue producing language or controlling bodily movements, for example. Because the borders of this type of tissue can shift spatially a little from person to person, surgeons stimulated the brain during the operation to test potential involvement in these functions. During this period of technical improvements, ethical standards advanced modestly. For example, Otfried Foerster (1931, 310), one of the leaders in this area, said: "Strong faradic [electrical] stimulation produces a convulsion, which may be limited to the eye muscles, but in other cases other movements occur." Clearly, the ethical concerns for patients left much to be desired. (Of course, in no way was the horrific treatment of animals, especially in the nineteenth century, acceptable either. The ethics of animal experimentation is a complex subject that continues to evolve.)

It was the neurosurgeon Wilder Penfield, together with his collaborators, who took the technique of electrical brain stimulation to the next level scientifically. Early in his career, Penfield was a surgical intern under Harvey Cushing himself. Because neurosurgery can be performed, as mentioned, under local anesthesia, patients can report sensations and feelings when stimulated in different parts. By exploiting the information thus garnered, Penfield and his colleagues generated detailed "functional maps" of the cortex across hundreds of patients summarizing the type of experience reported at each stimulation site. Their most famous discovery was the "homunculus" (Latin for "little person") first reported in 1937 and refined in 1950. The homunculus map is a grotesquely distorted outline of a body superimposed on a sketch of the top of the brain, depicting locations where stimulation produces sensory or motor effects (figure 6.2). The relative proportions of the body parts of the drawing (hand, foot, and so on) indicate

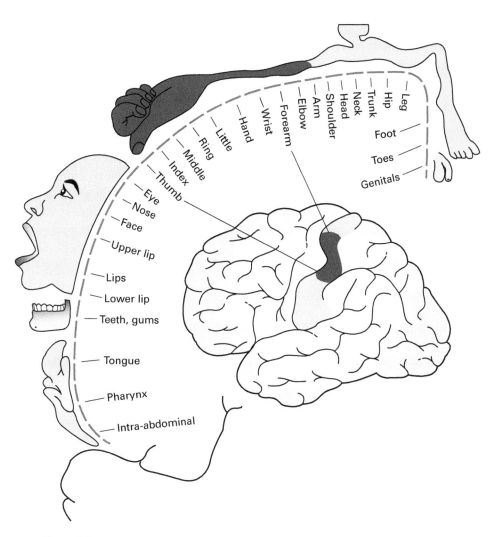

Figure 6.2

The so-called homunculus along the lateral surface of the somatosensory cortex in the parietal cortex. The parts corresponding to the foot, toes, and genitals are situated along the medial part of the brain, not visible from this side view.

the relative sizes of the regions whose stimulation influences the corresponding body parts. For example, the hands of the homunculus are much larger than the shoulder because the brain tissue where stimulation produces responses in the former is much larger than that where stimulation produces responses in the latter. And homuncular genitals are drawn next to the feet to indicate the relative positions of their cortical gray matter.

As we noted earlier, electrical stimulation of the cingulate cortex is, at times, accompanied by emotional experiences. But even before this type of evidence was obtained during neurosurgery, this cortical sector was among the first suggested to play a role in emotional processing, pretty much based on a hunch. In a landmark paper published in 1937, the neuroanatomist James Papez ventured that the cingulate gyrus was the cortical centerpiece of the "emotional brain," a cortical-subcortical circuit specialized for emotion. Whereas at the time there was little to indicate a link between the two, Papez cited medical cases of tumors in this region associated with "change in personality or character" and "loss of spontaneity in emotion" to back up his proposal (actually, not unlike Gage's purported changes, although his lesion was substantially more frontal, just behind the bones of the eyes). It was hardly solid evidence, given that clinical cases are rarely clean; more often than not, lesions are large or diffuse and damage multiple sites. So why did Papez choose the cingulate cortex as the cortical anchor of his emotion circuit?

In 1878, Paul Broca—the very same who examined Tan and localized language to the left frontal cortex (see chapter 4)—published a hugely influential manuscript with an unwieldy title: *Comparative anatomy of the cerebral circumvolutions: The great limbic lobe and the limbic fissure in the mammalian series.*[6] Broca proposed the existence of a "great cerebral cortical system" that encircles the limbus (or edge) of the hemispheres. At the broadest level, he subdivided the cortex into two components: the *great limbic lobe*, comprising the bulk of the medial surface of the cortex (essentially the cingulate cortex as shown in figure 6.1), and the rest of the cortex—the rest being all of the cortex that is visible from outside (frontal, parietal, occipital), plus the cortex of the insula (discussed in the next section). Broca believed that the brain of primates was qualitatively different from that of other animals because of the "predominance of the frontal lobe," as he stated. This frontal dominance was accompanied by another significant change: the atrophy of the olfactory system. What's more, these two changes were not accidental but reflected a "true correlation": the enlargement of the frontal lobe and the devolution of the great limbic lobe.

Although Broca's observations were anatomical in nature, they were intimately connected to his thinking about mental functions. For him, the sense of smell was a *bestial* one that required only slight "intellectual involvement" and relied on the limbic lobe, a cortical sector that ranked low in the cerebral totem pole. Intelligence gained supremacy over the bestial sense by the elaboration of the frontal lobe and the concomitant atrophy of the limbic sector. The impact of Broca's ideas were enormous, and one can venture that the concept of a "limbic brain" has been one of the most influential concepts in all of neuroscience. This was his second major scientific home run, the first being the paper about the localization of the language mental faculty. The upshot? After Broca, it was natural to link the cortex along the medial surface of the brain with emotion (often equated with the bestial or irrational side) and the "outer" parts, especially the front, with cognition (the rational side).

Let's return to Papez and his outline of an emotion circuit. Papez's proposal was further extended by Paul MacLean, whose ideas about the "emotional brain" reverberate to this day. In 1949, MacLean wrote a paper introducing the "visceral brain," which he dubbed the "limbic system" a few years later. The system was composed of the great limbic lobe of Broca (that is, the cingulate gyrus along the medial surface of the brain), together with select subcortical regions, of which the hypothalamus was given particular importance because of the work by Bard, Cannon, and their contemporaries (chapter 5). MacLean's limbic system established an emotional brain that was largely segregated from parts believed to support reason, echoing a dichotomy with a long history in Western thinking.

Some of the Functions of the Cingulate Cortex

In the 1940s, electrical stimulation studies in both primates and humans started to uncover that multiple sectors of the cortex—not only subcortex—elicited autonomic system changes. Striking changes in respiration, blood pressure, heart rate, and pupil dilation resulted when stimulating the cingulate gyrus. Changes in vocalization were also observed, which is notable given that such changes are present during emotional and motivational states in particular—think of the aggressive pants of a gorilla or the appeasing sounds of a chimpanzee; vocalization is invariably altered when humans are emotionally aroused, too.[7] The autonomic changes on engagement of the cingulate cortex are entirely consistent with its anatomical connectivity given that it projects to multiple structures outside the cortex that participate in autonomic processes. These pathways target the hypothalamus at the base of the forebrain, the periaqueductal gray (PAG), and other upper brainstem areas, as well as to structures in the medulla. The cingulate cortex thus influences multiple levels across the neuroaxis.

Indeed, the potential for the cingulate cortex to alter the state of the body is remarkable, as this cortical territory has the most extensive "descending" projections (those directed at non-cortical structures) of any other part of the cortex (figure 6.3).[8] Therefore, it is not surprising that this cortex is often viewed as an outflow, or motor, station. Nevertheless, the cingulate receives "ascending" signals from the brainstem. One of the most notable of these is from a nucleus in the medulla that is the major viscerosensory cell group in the brain. The area, called the nucleus of the solitary tract, receives inputs from the respiratory, cardiovascular, and gastrointestinal systems. Signals from pain-sensitive circuits also reach the cingulate through the thalamus. Overall, the cingulate cortex participates in two-way signal communication: not only does it participate in motor autonomic functions affecting the body, but it is sensitive to signals that convey the state of the body, too.

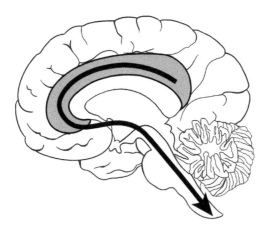

Figure 6.3
Descending connections from the cingulate cortex branch out at multiple levels, reaching subcortical structures at the base of the forebrain, midbrain, pons, and medulla.

Appraisal refers to the evaluation of an internal or external stimulus, and those that are significant induce an emotional reaction, the magnitude, duration, and quality of which result from the appraisal process. Functional MRI research indicates that responses in the cingulate cortex reflect appraisal.[9] For example, in studies in which participants were asked to rate the aversiveness of objects that were paired with mild shock, cingulate responses were positively correlated with participants' ratings; those that indicated that the object was more aversive exhibited stronger fMRI signals (figure 6.4).

(The use of mild shock may conjure thoughts of dreadful experiments performed by psychologists in the past, or even perhaps scenes from horror movies. However, modern experiments with mild electrical stimulation performed during functional MRI scanning [or outside of the scanner] employ stimuli that are well tolerated by the majority of participants. In studies in my laboratory, for example, participants determine their own level of stimulation by increasing and decreasing the intensity themselves, until attaining a level that is uncomfortable but not painful. Although the stimulus is clearly unpleasant, there is no other way to study negative emotion without participants experiencing something that is, well, unpleasant. But obviously this must be done in an ethical manner, with consenting adults. Participants are free to discontinue the experiment at any time. In well over a decade of performing such experiments, participants in my lab have stopped participation very few times.)

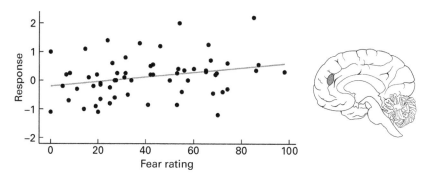

Figure 6.4
Ratings of aversiveness are correlated with responses in the cingulate cortex brain region (indicated in the line drawing of the brain). In the experiment, participants rated the aversiveness of objects previously paired with mild shock.

We saw in chapter 5 that the amygdala plays a major part in fear conditioning: The pairing of an initially neutral stimulus with an intrinsically aversive item leads to the once-neutral item acquiring negative properties. But what happens if the conditioned stimulus is no longer paired with the aversive item? For example, a sound that was paired with shock is now presented by itself. If this keeps on happening, it signals that the conditioned stimulus no longer predicts the unconditioned one. Naturally, it behooves the animal to learn that the stimulus is again neutral, a process called *extinction learning*. Interestingly, learning that a conditioned stimulus no longer forecasts an aversive event is not a passive, decay-like process of forgetting the previous fear memory. Instead, it is an active learning process in itself and needs to be sufficiently established for the animal to cease reacting to the neutral stimulus—which would be clearly maladaptive now that the stimulus is harmless. Extinction learning, which one can intuit has important implications for the understanding of clinical conditions such as anxiety, has been extensively studied. Several brain structures participate in this process, with the cingulate cortex being an important one. We'll return to extinction learning in a lot more detail in chapter 11.

The anterior part of the cingulate cortex is also linked to pain. Persistent pain is measured by means of self-report—a patient reports feeling back pain, say, and that it prevents her from playing tennis as she used to. Given the prevalence of pain-related conditions in the population at large, researchers have sought "objective markers" of this affliction—something akin to a blood test for a condition like diabetes. In a series of neuroimaging studies examining a large number of participants, Tor Wager and his collaborators (2013) found multiple brain regions, including the anterior cingulate cortex, whose signals are correlated with subjective discomfort levels: the higher the subjective level of pain, the stronger the response in this area.

These researchers sought to develop an objective measure of pain by using computational techniques from the field of machine learning. By way of analogy, consider the goal of recognizing images. After the algorithm is set up, an arbitrary image is provided as input and a label is generated as output, with the latter indicating the category of the input image—for example, "leopard" indicating the stimulus category. These algorithms are initially trained with a large number of sample images and then tested with novel ones. In what's called "supervised training," the machine learning algorithm is calibrated by providing images together with their known labels; the "supervisor" thus needs to know their identities.

Wager and colleagues adopted a similar approach to their functional MRI data.[10] In their experiment, a participant received noxious heat during the painful condition; as a control, at other times they experienced innocuous warmth. They trained their algorithm to associate responses across brain areas with the condition in question. Based only on the brain responses during a specific experimental condition, the machine learning method generated its own prediction ("painful" versus "non-painful"). Their technique performed very accurately and exhibited sensitivity and specificity of 94 percent or higher when data from novel participants were provided. (Sensitivity refers to the "true positive" rate—that is, deciding "painful" when the input is painful; specificity refers to the "true negative" rate, or deciding "non-painful" when the condition is non-painful.) In other words, when tested with data never seen by the machine learning algorithm, the method was able to guess the experimental condition by inspecting brain responses—a type of "brain reading" (figure 6.5). Even more impressively, the researchers could predict actual pain ratings; in this case, they trained their algorithm to estimate the continuous pain rating (on a scale of 0 to 8). Actual and predicted pain levels showed a strong match. When a participant rated the

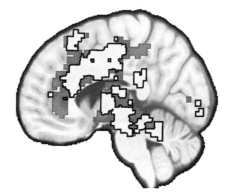

Figure 6.5
Contributions of brain areas to predicting pain. Stronger responses in areas indicated in light gray, including a large part of the cingulate cortex, predict pain states. They also predicted participants' numerical ratings of pain. Stronger responses in areas in darker gray predicted less pain.
Source: Image kindly provided by Tor Wager.

pain as higher intensity (closer to 8), the algorithm did so too; when they rated it as lower intensity (closer to 0), the method, on average, did the same. Overall, the approach by Wager and colleagues represents an exciting direction of research; it is rare to see such a quantitative angle in a field that is largely qualitative, even more so when targeted at something as subjective as how pain feels. No small feat.

Before moving to the next section, I'd like to discuss the enduring legacy of Papez's circuit and the limbic system idea, both of which tried to explain how emotion is organized in the brain. Although the term "limbic system" is probably one of the most broadly used in neuroscience, the concept has proved too unwieldy and unstable to be scientifically useful. (The terms is extremely popular in the general media, too; a search in the *New York Times* returned more than 200 hits.) Because agreement regarding the regions that belong to this system has never been attained, the term is used in a circular fashion to indicate the "emotional brain." As some have pointed out, "limbic system" substitutes naming for understanding.[11] Unfortunately, the term remains all too commonly employed by investigators, particularly those with more clinical or medical training. Indeed, it is somewhat baffling that medical texts describing the brain basis of emotion still discuss the limbic system in ways that go back to the original proposal by MacLean, if not all the way back to the circuit of Papez, although both of them reflect current knowledge rather poorly.

The Island of the Cortex

The external world impacts areas of the cortex that respond to visual, auditory, tactile, olfactory, and gustatory stimulation. As much as the world outside is rich in information, there is an equally luxuriant inner realm that pertains to the state of the body, including the soft internal organs, the viscera, and the body's outer layer, the skin, which is the largest organ of the body. A wide gamut of signals supports sensations related to temperature, pain, itching, tickling, sensual touch, muscular and visceral impressions, vasomotor flush (the sensation of sudden flushing and sweating), thirst, hunger, even "air hunger" (try holding your breath for 30 seconds). The cortex of the insula registers the state of the body in a precise and signal-specific fashion—it is a sophisticated sensory cortex (figure 6.6).

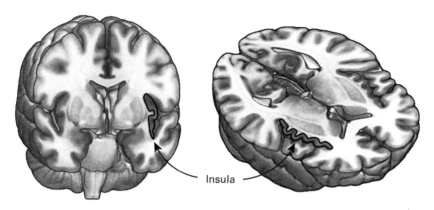

Figure 6.6

The cortex of the insula is visible when the brain is cut, thereby exposing the "internal cortical island."

Viscerosensory signals from the body reach the lower brainstem before eventually reaching this peculiar part of the cortex. Like an island, and covered by cortical tissue itself, the insula is not visible from the outside (if the brain were exposed and one were looking from the side). It is as if the cortical surface had decided to expand by creating a second cortical fold on top of the inner one (the insula); in fact, part of the cortex covering the insula is called "operculum," meaning "lid" in Latin.

Early hints that viscerosensory signals reach the cortex go back to the 1950s, when researchers stimulated the vagus, a large nerve exiting the brain that innervates the heart, the stomach, and intestines. When investigators stimulated the vagus nerve in monkeys and other animals, they noted activation of a "vagal receptive cortex" corresponding to the insular cortex. The role of the insula in the conscious appreciation of visceral sensation was vividly demonstrated in the studies by Penfield and colleagues, as they electrically stimulated the cortex of patients during neurosurgery.[12] As Penfield moved the electrode down along the primary sensory cortex, he identified a region extending just beyond the tongue in the homunculus, where electrical stimulation produced taste sensation. When he moved the electrode further into the insula, the patients reported oropharyngeal, esophageal, or even gastrointestinal sensation. Whereas the patients volunteered a variety of descriptions about their experiences, none of them reported emotional responses—they were more of a sensory nature. We now know that both sympathetic and parasympathetic bodily signals are conveyed to the insula. In fact, anterior insula responses reflect the internal state of the body along all of the dimensions outlined before, in a very real sense generating a map of "feelings from the body."[13]

We are in constant synergy with our surroundings, and touch is an important but often neglected component of such interactions. Touch helps acquire information about textures and shapes, aiding inferences about material properties and object identity. Touch also possesses an affective component; some experiences are pleasant (soft brush stroking, say) while others are unpleasant (stroking with sandpaper). Touch promotes affiliative, collaborative, and sexual behavior, too. Tactile social interactions even benefit mental and physical health. Lesion studies have demonstrated that the insula is important for different forms of touch sensation, including inferring sensory pleasure and emotion-related dimensions of the stimulation. And, for example, soft brush stroking produces functional MRI responses in the insular cortex (as well as in the somatosensory cortex; see figure 6.2).

By and large, researchers treat the insula as an autonomic sensory *input* station, in contrast to the suggested "motor," or output, autonomic role of the cingulate cortex. But, as we saw, the cingulate receives viscerosensory signals, too, so it participates in sensory processes in addition to participating in prominent output functions. The insula also contains descending projections that affect the body.[14] Thus, in both the cortex of the cingulate and the insula, *bidirectional* communication exists, albeit with asymmetrical efficacies.

In one of my early functional MRI studies, we sought to understand the mechanisms supporting good performance in a challenging cognitive task (Pessoa et al. 2002): How do participants maintain in mind for a few seconds a briefly presented visual pattern? Although the context was rather different, our approach was similar to the one employed by Wager and colleagues to study pain. Remember that, by using responses across brain regions, they attempted to predict if the participant was in a painful or non-painful state. In our case, the goal was to use brain signals to predict if the participant performed correctly or not in a given trial. Given that the task was effortful and stimuli were visual in nature, we weren't entirely surprised that signals in visual cortex and regions important for cognitively demanding tasks could be used to predict performance. In addition to these regions, we found that signals in the anterior insula closely correlated with task behavior. When discussing these findings, I remember that my colleagues and I were puzzled about this region's involvement during a cognitive task. Indeed, for several decades following the seminal studies of functions of the insular cortex, this cortical lobe was regarded largely as a viscerosensory part of the brain.

When MRI machines became more accessible worldwide in the late 1990s, researchers started observing responses in the insula during experimental conditions containing emotionally evocative stimuli, such as viewing pictures of a mutilated body. These responses were expected, as seeing such stimuli produce bodily sensations (imagine yourself viewing a picture of mutilation, perhaps from a medical text). But, gradually, functional MRI studies started to paint a different picture when responses in the insula were observed consistently in tasks spanning perceptual and cognitive conditions, much like I observed in the study of task performance.

During most functional MRI studies, the entire brain is scanned. So, even if an investigator has a pet theory that certain conditions evoke responses in particular brain regions, given that brain-wide data are collected, unexpected findings arise more readily. When more and more studies reveal a consistent pattern of results, the strength of the evidence grows accordingly. Indeed, when my colleagues and I performed the large-scale analysis of thousands of neuroimaging studies (discussed in chapter 4), the anterior insula was among the most *functionally diverse* regions of the entire brain (Anderson, Kinnison, and Pessoa 2013). Responses were observed across studies of perception, cognition, emotion, motivation, and action, suggesting that this area is much more than a dedicated viscerosensory sector—it participates in a very broad range of processes. To date, we still know little about *how* the insula contributes to such varied mental functions.

The Cortex above the Eyes

At age 35, patient EVR underwent neurosurgery to treat a meningioma, a slow-growing tumor that forms from the meninges, the membranes surrounding the brain.[15] Following recovery, EVR exhibited profound personality changes. His marriage had been stable and successful; after surgery, he divorced his wife of many years and within months remarried, but the marriage was short-lived. Previously with a keen sense of business, EVR attained considerable financial success; after his surgery, he entered into a series of brief but disastrous business ventures, one of which led him to bankruptcy. He had been an accomplished professional, securing promotions for good

performance; afterward, he was not able to maintain employment, ultimately having to live in a sheltered environment. You might think that
EVR was simply not "intelligent enough" anymore to be successful profes-
sionally, and that even his personal life was affected by this change. But
that doesn't explain his difficulties. After the surgery, he was an avid con-
versationalist that often came across as intelligent, charming, and witty. In
standardized tests, he scored more than two standard deviations above the
mean (IQ = 135, where the mean is centered at 100). So, he was skilled and
"intelligent enough" to hold a job.

The surgery on patient EVR excised a part of the frontal lobe called the
orbitofrontal cortex, which sits just above the orbits of the eyes and extends
back a few centimeters forming the frontal base of the brain (figure 6.7).
The personality alterations that he experienced are similar to those attributed
to Phineas Gage discussed earlier in the chapter. Although Gage's character
changes now seem to have been somewhat transient and largely inflated (at
some point, Gage actually became a successful long-distance stagecoach in
Chile), patients like EVR demonstrate that comparable "personality trans-
plants" can occur. Patients such as EVR do not exhibit drastic modification
of emotional behaviors; for example, they don't necessarily become severely
depressed, although they may experience anger more frequently than before.
Instead, the patients often become emotionally "shallow," experiencing an
overall dampening of affect. And in some cases they become callous to the
point of exhibiting what's called "pseudopsychopathy." Because the condition results from brain impairment, it has also been dubbed "acquired sociopathy" to highlight the fact that it results from the injury and to distinguish

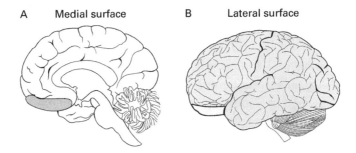

Figure 6.7
The orbitofrontal cortex seen from the middle (a) or side (b) of the brain. This cortex
sits just above the orbital bone and eyes and extends further inward into the brain.

it from conventional psychopathy, in which the cold-blooded traits emerge in childhood and adolescence with no gross structural brain changes.[16] Although there are many ways to define psychopathy, in broad terms, it's a personality disorder displaying persistent antisocial behavior, impaired empathy and remorse, and bold, disinhibited, and egotistical traits.

We discussed earlier in the chapter how the cingulate cortex and the insula are involved in autonomic processes. The same early wave of studies using electrical stimulation of cortical sites revealed the participation of the orbitofrontal cortex in these functions, too—for example, changes in respiration, blood pressure and heart rate, and pupil diameter. Today, much of the research on the orbitofrontal cortex aims at understanding how it contributes to the processing of rewards and, more generally, the computation of value. For instance, what is better for me now—to continue reading this book or to go out with some friends? We'll come back to these questions in chapter 10.

Coda

Early thinking in neuroscience was heavily influenced by the social and biological notions of Victorian England, where concepts such as progress, hierarchy, and control were much in vogue to justify the enrichment of a subset of society. In this context, the brain was viewed as comprised of "primitive" and "advanced" parts, with the subcortex and cortex existing as paradigmatic representatives of these two types of territory. Yet, by the 1950s, the participation of the prefrontal cortex in autonomic processing had been conclusively demonstrated. The crown jewel of "rational" processing was also bidirectionally involved in respiratory, cardiovascular, and even gastric mechanisms, all "lowly functions," including the lowest of them all. Every neuroscience textbook or book about the brain describes how parts of the cortex are important for handling the external world—vision in the occipital lobe, audition in the temporal lobe, somatosensation in the parietal lobe. But the cortex is equally important for taking care of the internal world of the body. Modern research emphasizes the involvement of the prefrontal cortex in the "highest functions"—planning, manipulation of information, prioritization of behaviors, and so on—to such an extent that its participation in monitoring and controlling the body is often forgotten. This view is shortsighted: complex behaviors involve a close interaction between the internal and external realms.

7   Cognition and the Prefrontal Cortex

We'll discuss two functions central to cognition: working memory and attention. Working memory has to do with how we keep information active while the information is useful for the actions and thoughts that we perform. Attention, as the common usage of the term implies, refers to what we monitor in the world, what really matters. Next, we discuss how these multifaceted functions engage the prefrontal cortex. Neuroscientists like to describe this part of the brain as it's crowning apex, where cognition—including abstract thought—reigns supreme. The chapter then picks this idea apart by tracing its origins and discussing how to think about the brain as a distributed system. We thus start building the case for an entangled view of the brain in terms of interacting parts.

Researchers have always sought to unravel features of the brain that are, presumably, uniquely human, or that at least confer the species with "superior" mental capabilities. The pseudoscientific phrenology movement led by Franz Gall and his disciples, discredited as it is now, galvanized the pursuit of localizing mental functions to brain territories. Phrenology was particularly active from the 1810s to 1840s and sought to find functions in the brain by observing and exploring the skull; proponents would run their fingertips and palms over a person's skull to feel for distinct patterns, such as enlargements or indentations. Perhaps because of its location at the front and top of the brain, Gall placed humans' "highest functions" in parts of the frontal lobe.[1] As neuroscience took shape as an active area of research in the second half of the nineteenth century, the question of localization received much attention and was greatly invigorated by Paul Broca's 1861 report linking language and the prefrontal cortex (chapter 4). His presentation before the *Société d'Anatomie* in Paris, which described his clinical and neurological conclusions about Tan, had a tremendous impact on both clinicians and experimentalists. Broca proposed that the frontal lobe was important for speaking and for higher intellectual functions, including judgment, reflection, and abstraction. Interest in this part of the brain was kindled considerably.

Seventeen years later, Broca would again have a major impact in shaping the prefrontal cortex's importance for intellectual functions. In his monumental theoretical paper of 1878 discussed in chapter 6, Broca developed this theme in a forceful manner, concluding that the primate brain can be distinguished from that of other mammals by "characteristics numerous and very striking." But chief among them was the *predominance* of the frontal lobe. Enamored with this view he threw all caution to the wind:[2]

> The simultaneous appearance of these numerous characteristics leads to major external changes and throws the entire cerebral morphology into such upheaval that one might believe oneself to be seeing a brand new order of things, as if the chain had been broken, nature had smashed its old molds, and the project had been started up again using a completely different set of plans.

In chapter 9, we will discuss brain evolution in more detail and explain how Broca, admittedly without the tools of modern neuroscience, painted an extreme—and incorrect—portrait of the organization of the mammalian brain. Yet, intellect and frontal cortex would be paired for the next 100 years—and still are.

Measuring Electrical Activity from Cortical Neurons

The same type of electrical stimulation studies that uncovered links between the cortex and the body (chapter 6), also sought to uncover the response properties of the cortex. But the discovery of the functions of particular brain parts was advanced enormously by the refinement of the electrophysiological recording technique in the late 1950s and early 1960s. With this method, electrodes were inserted into the cortex to directly record cellular activity, including spiking activity, the all-or-none firing of cells. Initially, recordings were made in anesthetized animals, but the technique was later extended to allow measurements in alert animals, although they still needed to be restrained. (Techniques to record neuronal activity in alert, freely moving animals are expanding rapidly.)

Early studies characterized responses in parts of the cortex that respond to sensory stimulation. Because the animals were restrained and anesthetized, this was a natural place to start. (You might wonder if cell firing can be detected when the animal is anesthetized; in the case of vision, even their eyes were kept open artificially. For sensory responses, basic response properties are similar when the animal is under anesthesia compared to alert.) A second key reason to study sensory regions was that it allowed researchers to choose and control stimuli rather precisely. For example, they could present a visual stimulus of a certain elongation, orientation, and speed at which it moved in front of the eyes. If it was an auditory stimulus, it could be a sound of precise intensity, duration, and wave characteristics.

Altogether, this strategy proved immensely productive and has kept neurophysiologists busy for decades. In the case of the visual system, little by little researchers built a body of knowledge describing how cortical areas process increasingly more complex objects. For example, in the primary visual cortex (area V1), cells respond to fairly specific and basic stimulus properties, such as the orientation of a "bar" (think of a pen). Based on physiological and anatomical properties, researchers gradually assembled a catalog of areas (unimaginatively, some were called V2, V3, and V4). Cell responses in each area are sensitive to different stimulus properties, including object shape, color, and motion. By the middle of the 1980s, around 15 regions in occipital, temporal, and parietal cortex were classified as "visual." Whereas the responses in some parts of the occipital lobe are fairly basic and tied to the physical quality of objects, responses in parts of temporal and parietal cortex are noticeably more abstract. Some even respond to human faces! Cells in the lower part of the temporal cortex fire intensely in response to pictures of faces and, in some cases, are quite selective in their responding; they might respond to pictures of a certain individual vigorously but much less to other faces or not at all to pictures without a face. One of the cells recorded in a 2005 study of an epilepsy patient with implanted electrodes seemed to have a clear preference for pictures of the actress Jennifer Aniston, leading to infelicitous media reports of the "Jennifer Aniston cell."[3] Of course, were that cell to die, the patient would not suddenly lose the concept or even the visual image of the actress. The investigators eventually found out that the neuron also fired in response to pictures of her former castmate, Lisa Kudrow, likely reflecting memory associations of the patient in question.

While charting visual areas in the monkey brain, it didn't take long for researchers to reach the frontal cortex, where some cells are heavily involved in the control of eye movements. Primates, in particular, are attuned to the shape and form of objects in the world, which rely on detailed visual processing at the location where the eye is fixating. As neurophysiology gradually moved to parts of the brain where responses are not directly tied to stimulus properties (it doesn't matter if the visual object is green or yellow, at least insofar as this isn't important for the task), the focus of research shifted to "cognitive variables." How does an animal remember a stimulus, decide what to pay attention to, or switch between tasks?

Holding Information in Mind

In the early 1970s, a major discovery opened the door to the understanding of how cognitive processes are realized in the brain. Psychologists had been interested in how humans and other animals actively maintain information for several decades. If given a list of words to remember, after a brief interval of 20 seconds, say, how many items can people recall? Can they remember both the items and their order? This type of "working memory" (which is different from long-term memory of events in the past) had been extensively investigated in monkeys as early as the 1930s and the importance of the prefrontal cortex already established: lesions of the so-called dorsolateral part of the prefrontal cortex substantially impair working memory abilities.

In a typical working memory experiment, on each trial, a visual stimulus is first shown. A delay period of several seconds then ensues during which the stimulus is absent (in early experiments before the use of computer screens, the monkeys' view was obstructed), followed by the presentation of the so-called test stimulus. The task is to indicate whether or not the test matches the initial stimulus. Successful performance thus necessitates some form of memory trace of the first display, which must be further matched to the test to determine the correct answer.

What type of neural signal could bridge the gap between the two visual items during the delay? Cell responses to visual stimuli are transient and decay back to pre-stimulation levels within half a second or so (in the absence of stimulation, many cells are not completely quiescent but respond at low "baseline" level). Thus, researchers were looking for evidence of *sustained* cell firing during the delay, and that is what they found when recording from neurons in the prefrontal cortex of rhesus monkeys (figure 7.1). The prolonged activity was interpreted to be the neural correlate of remembering

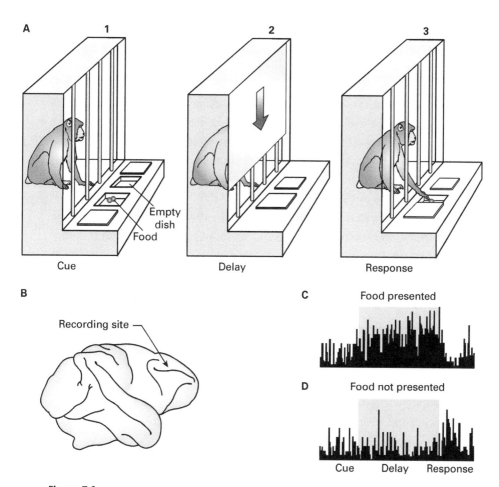

Figure 7.1
Keeping information in mind when it is out of sight. (a) Early experimental setup. (b, c, d) Cell recordings in the prefrontal cortex uncovered cells with sustained firing during the delay interval when the animal does not see the food item.

during the delay and taken to be the *neural signature* of working memory. Two independent papers published in 1971 discovered sustained firing and were enormously influential. For the first time, neuroscientists had shown that an abstract mental operation that wasn't directly linked to a physical stimulus—no stimulus is shown during the delay—could be identified. Studying cognitive tasks was thus viable, and the cellular mechanisms that underlie complex mental functions could be uncovered.

Can we conclude that cells with uninterrupted responses during the delay interval implement short-term remembering? For one, they could be tied to motor preparation, because at the end of the trial the monkey has to indicate whether or not the test stimulus matches the initial one—for example, by touching a response area. So, sustained firing might be related to the motor part of the task, not the mnemonic component. How can we strengthen the evidence in favor of maintenance processes? Cells in the prefrontal cortex prefer certain types of stimuli, of a certain shape or color, say. Multiple studies have shown that sustained responses are tuned to the same attributes of the to-be-remembered stimulus, which is precisely what would be required for a true neural signature (for example, Constantinidis, Franowicz, and Goldman-Rakic 2001). For example, if a prefrontal cell has a preference for a round stimulus when displayed briefly, it will fire vigorously during the delay if that very stimulus needs to be remembered (recall that no stimulus is shown during the delay); otherwise it will respond more weakly.

Should I Pay Attention to You?

Consider three lionesses as they attack a giraffe. They aim to confuse the prey and bounce its attention away; the giraffe cannot keep track of everything that is going on. Animals are not capable of processing all of the information they receive perfectly. How do they *select* some information while ignoring other information sources?

Individual cells in the visual cortex don't respond to stimuli everywhere; they respond only to objects in delimited parts of the field of vision, called the "receptive field." In addition, cells exhibit preference for particular features: Is it moving? Is it colored? And so on. In visual area V4, for example, neurons often exhibit preference for stimuli of certain colors. Researchers capitalized on these properties to investigate the mechanisms of visual attention.[4] How do the responses to something that matters differ from responses to an item that is less relevant? First, the researchers determined an "effective" stimulus that drove a cell's response vigorously, as well as an "ineffective" one that produced weak responses. They then placed both stimuli within the cell's receptive field (so both were capable of eliciting responses from the cell) and taught the animal to pay attention to one of them while ignoring the other. When they rewarded the animal for indicating the elongation (horizontal or vertical) of the relevant one, they found that the locus of attention had a dramatic effect on the cell's response. If the monkey paid attention to the effective stimulus, the cell responded vigorously. Now, when the monkey paid attention to the ineffective stimulus, the output was considerably decreased. This is remarkable because the actual physical stimulation is exactly the same in both conditions (figure 7.2). What changes is the *state* of the animal: In one case the effective stimulus is significant, and in the other the ineffective stimulus is the one that matters to obtain a reward by the experimenter. To a good extent, then, the responses were determined by the attributes of the attended object. It's as if the brain implemented a *filtering* mechanism capable of reducing the influence of the irrelevant information while zooming in on the important information at that time.

Since findings like those in figure 7.2 were first described in the mid-1980s, the literature on the mechanisms of attention has blossomed. We now know that the brain carries out a number of *competition* mechanisms: In the case of vision, the processing of certain objects or parts of the visual field are favored, whereas other objects or spatial locations are de-emphasized. Attention is better understood not as a single process but as a collection of processes that help solve this problem: How does the brain choose information that is relevant while ignoring less pertinent signals?

The Final Frontier: Frontal Cortex

We saw that the frontal cortex plays an important role in working memory—lesions impair a monkey's ability to actively maintain information in the mind. Attention is also believed to involve the frontal cortex. In the example described above, one of the regions influencing visual responses in V4 is an area in the frontal cortex called the "frontal eye field," which was originally known to control eye movements (hence the name) but contributes to attention, too. Anatomical pathways from the frontal eye field to the visual cortex allow the former to direct the latter so as to favor the processing of task-relevant information—the one the animal is instructed to pay attention to.

Before continuing, we need to clarify the distinction between the frontal and prefrontal cortex. The central sulcus is a prominent fissure on the lateral surface of the cortex that separates the parietal and frontal cortices. The first gyrus on the frontal side is where the primary motor cortex is found (figure 7.3); it's the motor part of the homunculus described by Wilder

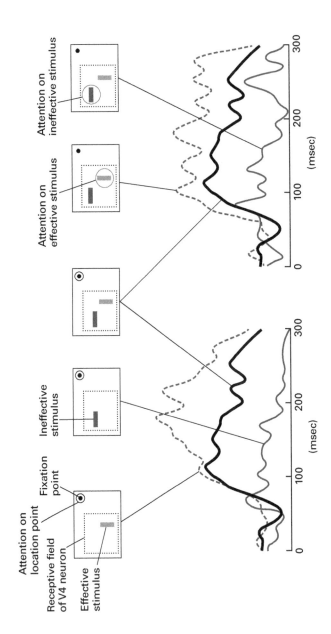

Figure 7.2

Attention can work as a filter helping to select relevant information. In the experimental setup, the animal is trained to keep the eye fixed during multiple stimulation conditions. When both effective and ineffective stimuli are in the receptive field, the response is intermediate (see middle rectangle and solid, dark gray line). When attention is focused on one of the stimuli while another stimulus is simultaneously present (two rightmost rectangles), the response approaches what is observed when a stimulus is presented alone (two leftmost rectangles). In other words, the process of attention appears to "filter out" the unattended stimulus.

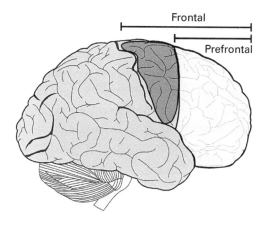

Figure 7.3
Frontal and prefrontal cortex. The prefrontal cortex includes areas that are less directly involved in motor function and are often considered involved in "abstract" processing.

Penfield (chapter 6), and cells there are intimately linked to motor actions. If we move further to the front of the brain, we encounter multiple "premotor" areas, where cell firing reflects motor intentions and planning but don't actually generate movements. To separate these regions of the frontal cortex involved in motor function from the more "noble parts" of the lobe (see discussion in the next section), the latter are described as "prefrontal."

Both working memory and attention are paradigmatic cognitive processes—they involve a prefrontal cortex that matches most neuroscientists' *expectations* of how the brain works.[5] But what is the general function of the prefrontal cortex? What does it do? Does damage to the prefrontal cortex "remove" intelligence from a person? Studies of patients with lesions have yielded somewhat of a puzzle or what's been described as the "riddle of the frontal lobe" (Damasio 1979). Indeed, it is frequently difficult to detect a neurological disorder based on patients' everyday behavior. They do not display obvious impairments in perceptual abilities, and their speech can be fluent and coherent. On conventional tests of intelligence, including standard IQ tests, they perform normally. But with more sensitive and specific tests, it becomes clear that frontal lesions disrupt function.

In some cases, the test administered is not so specific or even sophisticated. Take as an illustration (with no small amount of black humor) how the behavior of a prefrontal patient can be dominated by immediately available information. F. Lhermitte, a neurologist in Paris, directed patients to sit at a desk containing a hammer, a nail, and a picture (Lhermitte 1983). One patient picked up the items to hang the picture on the wall. In another setting, the neurologist approached, placed a hypodermic needle on the desk, dropped his trousers, and turned his back to the patient. What did the patient do? Unfazed, he simply picked up the needle and gave the doctor a healthy jab in the buttocks![6] If you are wondering what the patients were thinking, they would state something like, "You held out objects to me; I thought I had to use them."

As fascinating as they are, these examples don't help much in dissecting what might be happening. Is it a matter of disinhibition, only being able to focus on what is immediately present, loss of social knowledge, or some combination of these factors? Interestingly, in the cases above, the patients were capable of carrying out behaviors which, although relatively simple, were definitely nontrivial (think of the challenges of developing a robot to execute those tasks). So what does the prefrontal cortex do? More developed experimental paradigms converge on the idea that it is involved in what is called *cognitive control*. I'll illustrate this by describing two tasks: the Wisconsin card sort task and the Stroop task.[7]

In the Wisconsin card sort task, participants are asked to sort cards according to the shape, color, or the number of symbols (figure 7.4); they aren't told the rule by which to sort the cards, only whether or not they are correct after responding to each card. After the person obtains 10 correct choices, the rule switches. The catch is that the participant is not informed of this change, so after making a mistake, the person has to change the sorting rule. Once a correct response is generated, the rule is maintained until 10 correct choices are produced, and so on, until all cards on the pile are sorted. In a landmark study, the neuropsychologist Brenda Milner discovered that, in this task, all her frontal patients performed more poorly than her control participants, who had lesions of other parts of the cortex, such as the parietal and occipital cortex. In what way did they struggle? After the surgery, they tended to stick to one rule (sort by color, say) and to perseverate as they sorted the cards, *even* as they received continued negative feedback. Milner's approach was experimentally impeccable. Not only did she examine performance in patients with lesions elsewhere in the cortex, but each of her study groups was tested before and after the operation, allowing her to isolate the *change* in behavior based on lesion location.

Wisconsin card sort task

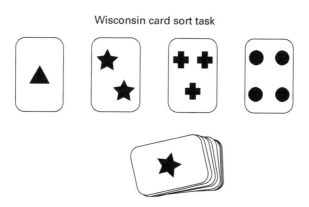

Figure 7.4
The Wisconsin card sort task requires sorting all cards according to shape, color, or number of symbols. Here, the test card at the bottom could match the first or second cards along the row above, based on number or shape, respectively. Participants are never explicitly told the rule, only if they are correct or not.

Nevertheless, interpreting the results of the Wisconsin task is challenging not least because it is difficult, and even some regular participants without a brain lesion perform poorly at it.

The Stroop task (named after its developer, J. Ridley Stroop) is attractive given its combination of simplicity and effectiveness. Imagine a page with words used for colors, such as "red," "green," and so on, written in black ink. Now imagine writing the words with pens of different colors, including one that matches the word and one that doesn't—for example, the word "red" written in red ink (called congruent) or the word "red" written in green ink (called incongruent). Your task is to tell the color of the ink used to write the word. If the word's name doesn't match the ink color, the task feels somewhat harder than it should be. In the Stroop task, the participant is asked to either read the word or to name the color in which it is written in different trials. Every stimulus contains two properties (word meaning and color), and the participant must attend selectively to the appropriate attribute to perform adequately. The key comparison is to contrast a person's reaction time to answer incongruent and congruent trials—one is a little slower in the former. Patients with frontal lobe damage have difficulty with this task, particularly in more challenging versions when the instructions are provided shortly before seeing the word (Dunbar and Sussman 1995).

The Stroop task is called a *conflict task* because during incongruent trials one is asked to focus on a particular dimension (color) in the presence of competing information (word meaning). Both *detecting* the presence of conflict and *resolving* it are important cognitive functions. Versions of conflict occur in many behavioral contexts, in particular when more automatic and controlled behaviors interact. When certain actions are practiced over and over, they become habitual, a process of *habit formation* that's important for making them efficient—that is, more automatic. Now, once the behavior is performed in a fairly automatic fashion, it can be rather difficult to break it: Once triggered, it's executed. But if a habitual action is uncalled for, it should be possible to recalibrate it, modify it, or call it off completely. That's when *cognitive control* comes into play. As a simple example, consider the act of stepping on the gas pedal when the light turns green at an intersection. For those of us who have been driving for many years, this action is completely habitual. But upon spotting a child crossing the street on a bicycle, one should be able to step on the brake pedal immediately (hopefully). Seeing both the green light and the child triggers the conflict (the detection part); releasing the accelerator and rapidly stepping on the brake solves the problem (the resolution part).

The Stroop task illustrates a fundamental aspect of cognitive control and *goal-directed* behaviors: the ability to select a weaker but task-relevant response (or source of information) in the face of competition from an otherwise stronger but task-irrelevant one.[8] Researchers believe that this is the central contribution of the prefrontal cortex, which allows adherence to goals in the presence of competing, stronger actions. It has been proposed that the key function of the prefrontal cortex is to enable and ensure adopting and following the "rules of the game." Think back to Lhermitte's patient: whatever the rules of being examined by a doctor in Paris, they don't include jabbing them in the buttocks with a syringe!

I'll end this section with a historical note. The quest to discover *the* function of the prefrontal cortex conceptualizes this territory as a unit, or at least as a relatively coherent functional entity. But the prefrontal cortex is large, and it is not surprising that the search to find its essence has been beleaguered with difficulties. By the 1980s, it was becoming increasingly clear that the heterogeneity of this large sector would need to be confronted for progress to accelerate. Even at a rather coarse level, we can distinguish three broad segments associated with the lateral, medial, and orbital surfaces.

The lateral aspect includes the dorsolateral prefrontal cortex, which is frequently linked to executive control functions. Both the medial and orbital surfaces have long been linked to emotional and motivational processing, too (chapter 6). Whereas most researchers have by now relinquished the goal of a more unified theory of prefrontal working and accepted its *multi-functional* nature, still, the territory is most frequently linked to cognitive processes, in particular the so-called *executive functions* required during non-routine, challenging situations. Later, in chapters 10 and 11, we'll explore how emotion and motivation engage the prefrontal cortex together with cognition.

From Sensation to Cognition, One Step at a Time

How are cognitive mechanisms built? In this section, I'll discuss a view that was popular for many decades, until at least the late 1980s. Although researchers currently don't subscribe to stronger incarnations of this model, the main idea lies dormant in the background, and current thinking is still influenced by it. It goes roughly like this.

Parts of the cortex are sensory, others are motor. In the former, cell responses are tied to stimulus and perceptual properties; in the latter, they are linked to movements and actions in a fairly direct manner. We can think of the prefrontal cortex as the part of the brain that's decoupled from immediate sensory and motor variables. In a nutshell, it deals with the *abstract*. On the sensory end, researchers think of the cortex in terms of information flowing from early sensory regions (for instance, visual, auditory) to parts of the prefrontal cortex (figure 7.5). Along the way, intermediate regions have cells whose responses are progressively more independent from sensory variables. Sensory signals provide a steady stream of signals that progress through a *cortical hierarchy* until reaching the prefrontal cortex. At each junction, responses become more refined—namely, less about physical properties. In the prefrontal cortex, they are sufficiently abstract that they can support "symbolic processing" of the type that possibly distinguishes humans from other apes, or maybe apes from other primates.

Consider the case of vision. Cells in the primary visual cortex (area V1) respond to basic stimulus properties, which are elaborated in subsequent areas, including V2, V3, and V4. After subsequent steps along the visual hierarchy, responses recorded along the temporal cortex reflect novel properties;

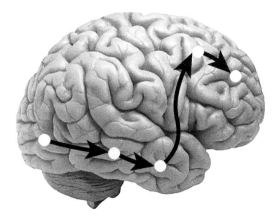

Figure 7.5
Flow of information processing in the brain. In the scheme favored by neuroscientists until at least the late 1980s, sequential processing steps move from the early visual cortex to the prefrontal cortex, where "abstract processing" is assumed to take place.

cells respond to object shapes and complex properties including "face-ness." Eventually, this processing stream reaches the prefrontal cortex. Within the prefrontal cortex itself multiple areas can be discerned, and as one moves toward the front of the brain, cells integrate multiple dimensions of the sensory world (they may respond to visual, auditory, and tactile properties simultaneously) in a manner that reflects the task at hand. Neurons care about features and objects that are behaviorally pertinent at that moment (for example, the color of a fruit matters when the animal is inspecting it for ripeness, not when navigating around it).

In this way, the information flows from sensory regions, which register the world "out there," to an *apex of integration* sitting atop the brain. It is thought that this hierarchical scheme protects the low-level stages, the ones doing basic sensory processing, so that they can reflect the environment with verisimilitude. The external world is thus represented with the least distortion possible. Interactions of diverse signals (such as from multiple sensory modalities) and integration with other processes (say, expectations and memory) are kept for intermediate and later stages, thus insulating the initial registering of the world from bias.

A related narrative can be told from the perspective of motor actions, but here we need to reverse the progression of signals. Signals flow from the regions of prefrontal cortex where the most abstract response properties are found, to so-called premotor regions in the frontal cortex engaged by action planning, and eventually to primary motor cortex, still located in the frontal cortex, where motor commands are issued. In other words, the sequence is from thought to musculoskeletal movements.

The Predictive Brain

While the conceptual framework outlined in the previous section was dominant, an alternative view was espoused by a minority of researchers, on and off, since the early days of brain research in the nineteenth century. Initially at least, it wasn't based on anatomical information per se but on conceptualizing the brain as a *prediction* device. In the leading view, action follows sensation in a progression that eventually culminates in a motor act: A monkey sees a fruit, registers its properties and position, and reaches out to grab it if it seems ripe. The predictive framework flips this logic on its head: Perception is directed by action so that effective behaviors can be generated. According to the traditional view, vision is relatively passive, like a camera pointed at the world, clicking away. In the predictive framework, vision is active and guided by endogenous computations that try to anticipate the most valuable future information for the animal.

From the standpoint of the active approach, the flow of information in the brain *can't* be like in figure 7.5. There should exist both connections from a "lower" to a "higher" area *and* the reverse—an idea that receives overwhelming empirical support. For instance, area V1 receives major projections from the visual part of the thalamus and is accordingly termed "primary" visual cortex. We know that V1 projects to area V2, too, but in addition, V2 projects to V1. We can therefore consider the V1 → V2 connections as "feedforward" and the V2 → V1 ones as "feedback." More generally, feedback connections, which are abundant in the cortex, provide a mechanism for predictions to influence earlier processing. In fact, Stephen Grossberg, a theoretical neuroscientist and a pioneer of the field of artificial neural networks (and one of my mentors in graduate school), proposed in the 1980s that feedforward and feedback pathways are arranged in what can be considered a basic building block: Connections from "lower" regions are reciprocated such that what ensues is a "consensus" between bottom-up and top-down signals (figure 7.6). This type of bidirectional architecture has profound implications for our understanding of how the brain works. Signals don't flow just one way but bidirectionally—higher areas project to, and influence, lower areas.

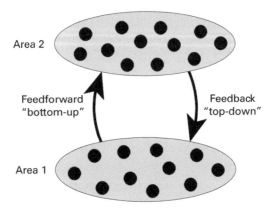

Figure 7.6
Bidirectional pathways between areas. The arrangement allows the two areas to mutually influence each other. One way to think about the pair is that one of them (here Area 2) provides "prediction signals" to the other.

The upshot is that brains are not passive. When a stimulus is processed, it does not encounter a tabula rasa. Instead, it is registered against a host of expectations constructed from prior experience, leading to the idea of a *matching* process between incoming information and feedback "template" signals. The template represents the system's predictions of the input and is updated to reflect the animal's past history.[9] Despite differences, predictive-brain approaches share a key concept: the brain doesn't reconstruct the external world but *constructs* a version of it.

Let's Do This All Together

In a critical assessment of the literature, the neuroanatomist Patricia Goldman-Rakic summarized the state of knowledge by the end of the 1980s as follows:

> The conclusion traditionally reached in virtually all comprehensive studies of cortical connections is that they are organized in a step-wise hierarchical sequence proceeding from relatively raw sensory input at the primary sensory cortices through successive stages of intramodality [that is, specific to one of the senses] elaboration allowing progressively more complex discriminations of the features of a particular stimulus. (Goldman-Rakic 1988, 146)

This statement is, of course, aligned with the traditional feedforward framework described in this chapter. But the time was ripe for different types of organization to be entertained. For one, computer science was evolving rapidly, and the centralized hardware organization of computers was changing to allow distributed processing. Instead of the architecture with a central processing unit, or CPU, parallel computers were designed with a large number of processing elements (some machines had thousands of them). A second impetus came from the field of artificial neural networks, which resurfaced quite strongly at the time. Although neural network models containing large numbers of artificial neurons—which could be thought of as simple processing units—were developed in the 1940s and studied by a small but very active group of investigators, the field of "connectionism," as some called it, only flourished in the mid-1980s. Neuroscientists took notice and started to think in more flexible ways.

Viewing the brain as a distributed system emerged as a guiding principle, as persuasively described by Goldman-Rakic in the paper quoted above. In a centralized architecture, signals converge at a specific element (the CPU, say), where they are further elaborated. Conceptually, this is the most straightforward type of arrangement one can imagine. Given that signals are all present at one location, they can be manipulated and operated on to generate an output. The distributed architecture, instead, carries its operations in a spatially scattered manner—and even the "results" themselves may be decentralized (figure 7.7). To illustrate with an early example, the neurologist Marsel Mesulam proposed that the parietal, frontal, and cingulate cortex work together to implement "attentional functions" (Mesulam

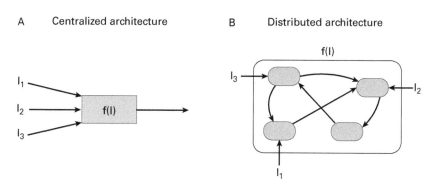

Figure 7.7
Centralized vs. distributed computation. (a) In a centralized system, a function $f()$ is computed by a processing element based on its inputs, generating an output. (b) In a distributed system, multiple basic processing elements interact such that the function $f()$ is carried out by the system as a whole.

1981). That is to say, mental functions that we call "attention" rely on joint contributions from regions within these three cortical sectors to carry them out. What's more, although working together, they provide distinct contributions to attentional processing, with the parietal, frontal, and cingulate cortices most important for sensory, motor, and motivation, respectively. Thus, while composed of multiple parts, the three areas of the attention circuit aren't exchangeable.

Coda

Let's recap the path taken in this chapter. Our goal was to describe some of the mechanisms that support cognition, with a particular focus on executive functions. Cognition reflects processing that is less tied to the sensory world. Distinct dimensions of the same stimulus can be relevant based on the context at hand, and the system needs to be updated on the fly, always at the ready. Simplistically, we can think of sensory signals as flowing from the periphery to the central headquarters (the prefrontal cortex), where information is manipulated and put together in complex ways. This transformation of perception into cognition is supported by an anatomical architecture that takes signals, step by step, from the sensory cortex to the prefrontal cortex. This view was indeed favored by most neuroscientists until at least the middle of the 1980s.

But another mode of communication, one based on feedback connections, must be considered, too. Abundant *bidirectional* connectivity fosters a view that processing is as much about exogenous as about endogenous signals, leading to an active, predictive system. Furthermore, parallel pathways are capable of conveying information in a distributed manner, creating an elaborate anatomical infrastructure that can support nonsequential and decentralized mechanisms. Thus, cognition is the product of much richer and nuanced mechanisms than a piecewise building-up process. In fact, in chapter 10, we'll describe how large-scale cortical-subcortical interactions are essential for constructing cognition and melding it together with emotion and motivation. Before doing so, let's discuss the concept of complex systems in chapter 8. To understand how the central nervous system supports sophisticated mental functions, we need to have a better grasp of *systems thinking*.

8   Complex Systems: The Science of Interacting Parts

What kind of object is the brain? The central premise of this book is that it cannot be neatly decomposed into a set of parts so that each one can be understood on its own. Instead, the brain is a highly entangled system that needs to be understood differently. The language that is required is the one of complex systems, which we now describe in intuitive terms. Whereas mathematics is needed to formalize it, illustration of the central concepts provides the reader with "intuition pumps."[1] Thinking in terms of complex systems frees us from the shackles of linear thinking, enabling explanations built with "collective computations" that elude simplistic narratives.

Kelp Carpets

Many coastal environments are inhabited by a great variety of algae, including a brown seaweed called kelp.[2] The distribution of kelp can be very uneven, with abundance in some places and a near absence in others. Ecologists noticed that in some coastal communities with tide pools and shallow waters largely devoid of kelp and other algae, killer whales (also called orcas) are also plentiful. The orcas don't eat kelp, so the negative relationship between the two must be purely incidental, right? Sea otters are a frequent member of coastal habitats, too, and their population has rebounded strongly since they gained protected status in 1911, just at a time when they numbered only 2,000 worldwide. Could they be the ones responsible for the lack of kelp in some areas? Sea otters don't consume kelp either, so what could be going on? It turns out that sea urchins are one of the most prevalent grazers of algae and kelp. And otters snack on sea urchins in large amounts. Therefore, because the presence of otters suppresses the urchin population, they have a direct impact on the kelp carpeting along the coast: The more otters, the fewer the urchins, and the richer the kelp. We see that otters and kelp are linked by a double negative logic: if you suppress a suppressor, the net effect is an *increase* (of kelp in this case).

But how about the orcas? The preferred meals of killer whales are sea lions and other whales, which are larger and richer in the fat content they need. But in some places these favored foods have become scarce. And with the large increase of otters given conservation efforts, the orcas appear to have turned to them as replacement meals. Altogether, we have a four-way relationship: ▲ killer whales → ▼ otters → ▲ urchins → ▼ kelp. This *cascade* is not a one-of-a-kind illustration of indirect effects; it is at the core of how ecological systems function. In other words, complex webs of interrelationships with many indirect effects—in fact, with multiple-step-removed indirect effects—are pretty much the norm.

Bacterial Decision Making

Bacteria are extremely simple creatures.[3] But when they are grown in a medium containing glucose and carbon dioxide, they can make all 20 kinds of amino acids, which are the building blocks of proteins, which do a lot of the heavy lifting in living things. When a specific amino acid is added to a glob of bacteria that is generating amino acids, the biosynthesis of that specific amino acid stops soon thereafter. But how do bacteria know that they no longer need to synthesize that particular amino acid?

In the 1950s, biochemists started to understand that amino acids are manufactured in several steps, starting with an initial "precursor" that is modified by a series of reactions leading to the amino acid. We can represent this in the following way: $P \rightarrow I_1 \rightarrow I_2 \rightarrow \ldots \rightarrow$ amino acid, where P stands for the chemical precursor and I for the various intermediate products. They also discovered that introducing a certain amino acid (call it amino acid A) could terminate the synthesis of *other* amino acids, indicating that amino acid A was having a negative effect somewhere along a synthesis pathway like the one above. What was more interesting, though, was that providing the amino acid isoleucine inhibited the production of isoleucine itself. If the system is producing a specific amino acid (isoleucine), one could imagine that adding more of it would further increase the overall amount of this compound. But just the opposite was found. Thus, the system automatically adjusted itself to prevent the overproduction of isoleucine.

This is an example of *negative feedback regulation*. Feedback seems simple enough, thermostats and autopilot systems being common in the modern world. However, it carries the core of a fundamental property: the ability of a system, even if very basic, to regulate itself. In its simplest form the idea is rather benign and unproblematic. But feedback muddies our intuitions about *causation*.

Consider again a system without feedback—let's call it SYSTEM 1 (figure 8.1). A precursor P causes some intermediate product I, which produces another chemical, and so on, until the last intermediate in the chain produces A. As far as causation goes, SYSTEM 1 is straightforward. Now, consider SYSTEM 2, which includes negative feedback. Here, if there's too much of A, the production of A itself will be inhibited so that its concentration will not increase further. SYSTEM 2 is not hard to understand, but with the small change (A loops back on the system), A has a causal effect on *itself*, which

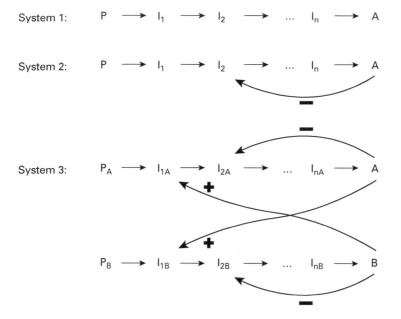

Figure 8.1

Feedforward, feedback, and interacting systems. System 1 is purely feedforward, while system 2 contains negative feedback. System 3 contains two chains: one that produces A, one that produces B. Each of them contains negative feedback. The chains are also positively coupled so that production of A encourages production of B, and vice versa.

means that A is both an *effect* (A is produced by the system) and a *cause* (A affects itself). Let's up things now and consider SYSTEM 3, which has feedback loops and interactions between two amino-acid pathways. Here, amino acids A and B affect and reinforce each other; production of A stimulates the growth of B, and production of B stimulates the growth of A. But production and growth of A and B are not unbounded because they have negative loops within their respective systems. Clearly, mapping out the mechanisms of production of the amino acids in SYSTEM 3 is substantially more challenging than the other two examples.

There is nothing odd, or potentially mysterious, about SYSTEM 3. All the functioning is mechanistic, in the sense that all parts function according to the standard rules of chemistry and physics. All that is present in *interdependence*. Why do we even need to bring this up? In many experimental disciplines, researchers are trained to think about causation pathways like that of SYSTEM 1 and, to some extent, SYSTEM 2. Thus, causation appears to work in relatively simple ways; for example, higher levels of cholesterol "cause" (with multiple intermediate steps) greater heart-disease problems. As elaborated in this chapter, however, systems like SYSTEM 3 can exhibit "complex" behaviors and "emergent" properties that are *qualitatively* different from those seen in simpler cases. And if the systems studied in biology are heavily interdependent, the field needs a change in perspective to move forward.

Of Predators and Prey

The Italian biologist Umberto D'Ancona was a prolific scientist who published over 300 papers and described numerous species. While studying fish catches in the Adriatic Sea, he noticed that the abundance of certain species increased markedly during the years of World War I, a time when fishing intensity reduced because of the war.[4] Puzzled by the observation, D'Ancona discussed it with the Italian mathematician and physicist Vito Volterra, who had become interested in mathematical biology (D'Ancona happened to be courting Volterra's daughter and, incidentally, the two would later marry). It is worth pointing out that when D'Ancona made his observations, ecology was not yet a systematic field of study (Charles Elton's now-classic *Animal Ecology* was published in 1927).

In the early 1920s, Volterra, and independently Alfred Lotka, mathematically described how interactions between a predator and its prey could be precisely written out (in Volterra's case, the prey being fish and the predator being fishermen). While we don't need to concern ourselves with the equations here, the model specifies that the number of predators, y, decays in the absence of prey and increases based on the rate at which they consume prey. At the same time, the number of prey, x, grows if left unchecked and decays given the rate at which it is preyed upon. The key point here is that x depends on y and, conversely, y depends on x. This interdependence means that we can eschew a description in terms of simple causation (say, "predation causes prey numbers to fall") and consider the predator-prey system as a *unit*. Put differently, predator and prey numbers co-evolve, and as such, characterizing and understanding them implies studying the "system" of predator plus prey.

By doing so, we aren't saying that there are no causal interactions taking place. Fishermen do kill fish and have an immediate impact on their population. But we can treat the predator-plus-prey pair as the object of interest. Whereas this is a relatively minor conceptual maneuver in this case, it will prove instrumental when a larger constellation of actors interacts.

Against Reductionism

The Lotka-Volterra predator-prey model formalized the relationship between a single predator species and a single prey species. Of course, natural habitats are not confined to two species; as the killer whale and kelp example illustrated, multispecies interactions are the norm. Thus, unraveling an entire set of interconnections is required for deeper understanding.

The prevailing modus operandi of science can be summarized as follows (Von Bertalanffy 1950, 134): "Explain phenomena by reducing them to an interplay of elementary units which could be investigated independently of each other." This *reductionistic* approach reached its zenith, perhaps, with the success of chemistry and particle physics in the twentieth century. In the present century, its power is clearly evidenced by dramatic progress in molecular biology and genetics. At its root, this attitude to science "resolves all natural phenomena into a play of elementary units, the characteristics of which remain unaltered whether they are investigated in isolation or in a complex" (Von Bertalanffy 1950, 134).

In the 1940s and 1950s, "systems thinking" started to offer an alternative mental springboard. Scholars surmised that many objects of study could be studied in terms of collections of interacting parts, an approach that could be applied to physical, biological, and even social problems. The framework developed, which some called *complex systems theory*, doesn't challenge the status and role of "elementary" units (no one was about to rescind Nobel prizes such as Ernest Rutherford's for the atomic model!). Again, in the words of one of its chief proponents, Ludwig Von Bertalanffy, the approach asserts "the necessity of investigating not only parts but also relations of organization resulting from a dynamic interaction and manifesting themselves by the difference in behavior of parts in isolation and in the whole organism" (Von Bertalanffy 1950, 135).

What does it mean to talk about "difference in behavior of parts in isolation and in the whole organism"?[5] Enter *emergence*, a term originally coined in the 1870s to describe instances in chemistry and physiology where new and unpredictable properties appear that aren't clearly ascribable to the elements from which they arise. When amino acids organize themselves—that is, *self-organize*—into a protein, the protein can carry out enzymatic functions that the amino acids on their own cannot. More importantly, they behave differently as part of the protein than they would on their own. But it's actually more than that. The dynamics of the system (that is, the protein) *closes off* some of the behaviors that would be open to the components (amino acids) were they not captured by the overall system. Once folded up into a protein, the amino acids find their activity regulated—they behave differently. Thus, one definition of emergence is as follows: a property that is observed when multiple elements interact that is *not* present at the level of the elements. Accordingly, it becomes meaningful to talk about two *levels* of description: a lower level of elements and a higher level of the system.

The growth of the complex systems approach was quickly popularized by expressions such as "system," "gestalt," "organism," "wholeness," and of course the much-used "the whole is more than the sum of its parts." In a manner that anticipated debates that would persist for decades, and still do, Von Bertalanffy stated as early as 1950 that "these concepts have often been misused, and they are of a vague and somewhat mystical character" (Von Bertalanffy 1950, 142). Even more presciently, he said that the "exact scientist therefore is inclined to look at these conceptions with justified mistrust."

Consider research in biology. The stunning developments of molecular biology, for one, raise the hope that *all* seemingly emergent properties can eventually be "explained away" and thereby deduced from lower-level characteristics and laws—the "higher" level can therefore be *reduced* to the "lower" level. Reduction to basic physics and chemistry becomes, then, the ultimate goal of scientific explanations. In this view, emergence is relegated to a sort of "promissory reductionism"—if not outright discredited—given that at a more advanced stage of science, emergent properties will be entirely captured by lower-level properties and laws. No doubt, it is extremely hard to argue against this line of argument. As the philosopher Terrence Deacon nicely states, looking at the world in terms of constituent parts of larger entities seems like an "unimpeachable methodology." It is as old as the pre-Socratic Greek thinkers and remains almost an "axiom of modern science."[6]

Both scientifically and philosophically speaking, the friction caused by the idea of emergence arises because it is actually unclear what precisely emerges. For example, what is it about amino acids as part of proteins that differs from free-floating ones? The question revolves around the exact status of "emergent properties." Philosophers formalize the terms used by talking about the *ontological* status of emergence—that is, concerning the proper *existence* of the higher-level properties. Do emergent properties point to the existence of *new* laws that are not present at the lower level? Is something fundamentally irreducible at stake? These questions are so daunting that they remain by and large unsolved—and subject to vigorous intellectual battles.

Fortunately, we don't need to crack the problem here and can instead use lower and higher levels pragmatically when they are *epistemically* useful—when the theoretical stance advances knowledge. To provide an oversimplified example, we don't need to worry about the status between quarks and aerodynamics. Massive airplanes are of course made of matter, agglomerations of elementary particles such as quarks (when put together, quarks form things like protons and neutrons, which are the stable components of atomic nuclei). But when engineers design a new airplane, they consider the laws of aerodynamics, the study of the motion of air, and particularly the behavior of a solid object, such an airplane wing, in air—they need no training at all in particle physics! So there's no need to really agonize about the "true" relationship between aerodynamics and particle physics. The practical thing to do is simply to study aerodynamics.

One could object to the example above because the inherent levels of particle physics and aerodynamics are far removed (figure 8.2); one level is too "micro" and the other is too "macro." More interesting cases present themselves when the constituent parts and the higher-level objects are closer to each other. For example, consider the behavior of an individual

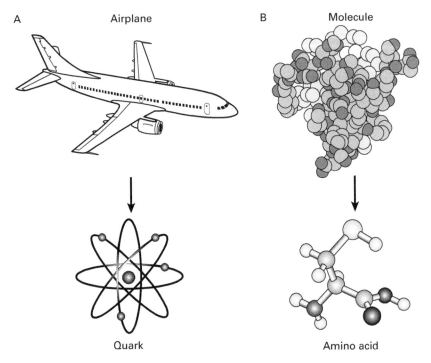

Figure 8.2
Levels of explanation and scientific reduction. (a) Describing airplane aerodynamics in
terms of elementary particles such as quarks is clearly not very useful. (b) Molecular
configurations in three dimensions may be investigated by determining how their
properties depend on chemical interactions between amino acids (which themselves
determine protein structure).

ant and the collective behavior of the ant colony, or the flight of a pelican
and the V-shape pattern of the flock. And of course, amino acids and proteins. As the researcher Alicia Juarrero says, it is particularly intriguing
when purely "deterministic systems exhibit organized and apparently novel
properties, seemingly emergent characteristics that should be predictable in
principle, but are not in fact" (Juarrero 1999, 6). And it's all the more fascinating when the systems involved are made of *very* simple parts that obey
straightforward rules. Understanding higher-level properties without having to solve the ontological question—are these properties *truly* new?—is
clearly beneficial.

We encountered John von Neumann previously in chapter 2. He not
only was one of the major players in defining computer science as we know

it, but his contributions to mathematics and physics are astounding. For instance, in 1932, he was the first to establish a rigorous mathematical framework for quantum mechanics. One of his smaller contributions was the invention of *cellular automata*, and this without the aid of computers, just with pencil and paper. (Another "minor" contribution by von Neumann was the invention of game theory, which is the study of mathematical models of conflict and cooperation.) A simple way to think of cellular automata is to imagine a piece of paper on which a regular grid is drawn. Each "cell" of the grid can be in one of two states: active or inactive, or 0 or 1 (think of a computer *bit*). The cells change state according to simple but precise rules, depending on the state of the cells' neighborhood. Different types of spatial neighborhood arrangements can be used, but for our purposes we will consider the simplest case, with just the cells to the left and to the right of a reference cell. A rule could turn the center cell active if either neighboring cell is active (called the OR rule); another rule could turn the center on if both neighbors are active (called the AND rule). If the cells start at some state—for instance, a random configuration of 0s and 1s—one can let them change states according to a specific set of rules and observe the overall behaviors that ensue (imagine a screen with pixels turning on and off). Remarkably, even simple cellular automata can exhibit rather complex behaviors, including the formation of hierarchically organized patterns that fluctuate periodically.

Although cellular automata were not widely known outside computer science circles, the idea was popularized more broadly with the invention of the Game of Life (or simply, Life). The game has attracted much interest not least because of the surprising ways in which patterns can evolve. From a relatively simple set of rules, some of the observed patterns are reminiscent of the rise, fall, and alterations of a society of living organisms and have been used to illustrate the notion that "design" and "organization" can emerge in the absence of a designer.

The examples provided by cellular automata, and others discussed in this chapter, suggest that we can adopt a pragmatic stance regarding the "true" standing of emergence. We can remain agnostic about the status itself but adopt a complex systems framework to advance the understanding of objects with many interacting elements. Let's discuss some ways in which this viewpoint is taking place in the field of ecology, the research area that originated the predator-prey models of Lotka and Volterra.

How Do Species Interact?

Ecology is the scientific study of interactions between organisms and their environment. A major topic of interest centers around the cooperation and competition between species. One may conjure investigators withstanding the blazing tropical sun to study biodiversity in the Amazon or harsh artic winters to study fluctuations in the population of polar bears. Although such field work is necessary to gather data, theoretical work is equally needed.

What are the mechanisms of species coexistence?[7] And how does the enormous diversity of species seen in nature persist, despite differences in the ability to compete for survival? Diversity indeed. For example, a 25-hectare plot in the Amazon rainforest contains more than 1,000 tropical tree species. As we have seen, in the 1920s, mathematical tools to model the dynamics of predator-prey systems were developed. The equations for these systems were further extended and refined in the subsequent decades and continue to be the object of much research. The study of species coexistence focuses almost exclusively on pairs of competitors so that when considering large groups of plants or animals, the strategy is to look at all possible couples. For example, one studies three pairs when three species are involved, or six pairs when four species are considered, or more generally, $n(n-1)/2$ interactions between n species. Do we lose anything when examining only pairwise interactions? *Higher-order interactions* are missed, as when the effect of one competitor on another depends on the population density of a third species or an even larger number of them. For example, the interaction between cheetahs and gazelles might be affected by hyenas, as hyenas can easily challenge the relatively scrawny cheetahs after the kill, especially when not alone (figure 8.3).

The importance of high-order effects is that, at times, they make predictions that diverge from what would be expected from only pairwise interactions. In a classic paper from 1972, entitled "Will a Large Complex System Be Stable?," the theoretical biologist Robert May showed formally that community diversity destabilizes ecological systems. In other words, diverse communities lead to instabilities such as the local elimination of certain species. Recent theoretical results show, however, that higher-order interactions can cause communities with greater diversity to be *more* stable than their species-poor counterparts, contrary to classic theory that is based on pairwise interactions.[8] These results illustrate that to understand a complex system (diverse community) of interacting players (species), we must

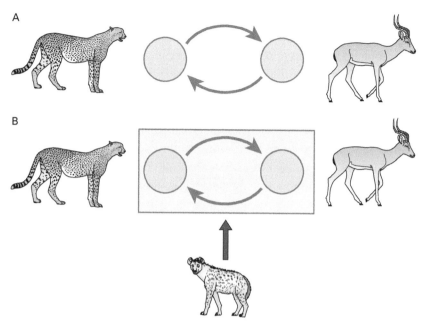

Figure 8.3
Species interactions. (a) Two-way interaction can occur, such as between predator and prey. (b) A higher-order interaction occurs when an additional element affects the way the two-way interaction behaves.

determine (emergent) properties at the collective level (including coexistence and biodiversity). Not only do we need to consider interactions, but we need to describe them richly enough for collective properties to be unraveled.

Neural Networks

Ideas about complex systems and the closely related movement of *cybernetics* didn't take long to start influencing thinking about the brain. For example, W. Ross Ashby outlined in his 1952 book *Design for a Brain* the importance of *stability*. Cybernetics researchers were interested in how systems regulate themselves and avoid instability. In particular, when a system is perturbed from its current state, how does it automatically adapt its configuration to minimize the effects of such disturbances? Not long afterward, the field of *artificial neural networks* (or simply neural networks) started to materialize. The growth of this new area proceeded in parallel with "standard" artificial intelligence (AI). Whereas AI sought to design intelligent algorithms by capitalizing on the power of newly developed computers, neural networks looked at the brain for inspiration. The general philosophy was simple: Collections of simple processing elements, or "neurons," when arranged in particular configurations called architectures generate sophisticated behaviors. And by specifying how the connections between artificial neurons change across time, neural networks learn new behaviors.

Many types of architecture were investigated, including purely feedforward and recurrent networks. In feedforward networks, information flows from an input layer of neurons, where the input (for instance, an image) is registered, to one or more intermediate layers, eventually reaching an output layer, where the output is coded (indicating that the input image is, say, a face). Recurrent networks, where connections can be both feedforward and feedback, are more interesting in the context of complex systems. In this type of organization, at least some connections are bidirectional and the systems can exhibit a range of properties. For example, *competition* can occur between parts of the network, with the consequent suppression of some kinds of activity and the enhancement of others. Interested in this type of competitive process, in the 1980s, Stephen Grossberg, whom we mentioned in chapter 7, developed Adaptive Resonance Theory (Grossberg 2021). In the theory, a *resonance* is a dynamical state during which neuronal firings across a network are amplified and synchronized when they interact bidirectionally—they mutually support each other (see figure 7.6 and accompanying text). Based on the continued development of the theory in the decades since its proposal, these types of bidirectional, competitive interactions have been used to explain a large number of experimental findings across the areas of perception, cognition, and emotion, for example.

Nonlinear Dynamical Systems

As we've seen, in the second half of the twentieth century, complex systems thinking began to flourish and influence multiple scientific disciplines, from the social to the biological. The ideas gained considerable momentum with the development of an area of mathematics called *nonlinear dynamical systems*. It is no exaggeration to say that nonlinear dynamical systems provide a language for complex systems. This branch of mathematics studies techniques that allow applied scientists to describe how objects *change* in time. It all started with the discovery of differential and integral *calculus* by Isaac Newton and Gottfried Leibniz in the last decades of the seventeenth century. Calculus is the first monumental achievement of modern mathematics, and many consider it the greatest advance in exact thinking. Newton, for one, was interested in planetary motion and used calculus to describe the trajectories of planets in orbit.[9]

Research in dynamical systems revealed that even putatively simple systems can exhibit very rich behaviors. At first, this was rather surprising because mathematicians and applied scientists alike believed deterministic systems behave in a fairly predictable manner. Because of this intuition, many techniques relied on "linearization"—that is, considering a system to be approximately linear (at least when small perturbations are involved). What is a linear system? In essence, it is one that produces an output by summing its inputs: The more the input, the more the output, and in exact proportion to the inputs. Systems like this are predictable and thus stable, which is desirable when we design a system. When you change the setting on the ceiling fan to "2," it moves faster than at "1"; when set to "3," you don't want it spinning out of control all of sudden!

The field of nonlinear systems tells us that "linear thinking" is just not enough. Approximating the behavior of objects by using linear systems does not do justice to the complexity of behaviors observed in real situations, as is most clearly demonstrated by a property called *chaos*. Confusingly, "chaos" does not refer to erratic or random behavior; instead, it refers to a property of systems that follow precise deterministic laws but *appear* to behave randomly. Although the precise definition of "chaos" is mathematical, we can think of it as describing complex, recurring, yet not exactly repeatable behaviors. (Imagine a leaf floating in a stream caught between rocks and circling around them in a way that is both repeating but not identical.) The theoretical developments in nonlinear dynamics were extremely influential because, until the 1960s, even mathematicians and physicists thought of dynamics in relatively simple terms.[10]

The field of dynamical systems has greatly enriched our understanding of natural and artificial systems. Even those with relatively simple descriptions can exhibit behaviors that are not possible to predict with confidence. Nonlinear dynamical systems not only contribute to our view of how interacting elements behave, but they define both a language and a formal system to characterize "emergent" behaviors. In a very real sense, they have greatly helped demystify some of the vague notions described in the early days of systems thinking. We now have a precise way to tackle the question of "the sum is greater than its parts."

The Brain as a Complex System

Complex systems are now a sprawling area encompassing applied and theoretical research. The goal of this chapter is to introduce the reader to some of its central ideas (a rather optimistic proposition without writing an entire book!). Whereas the science of complexity has evolved enormously in the past 70-odd years, experimental scientists are all-too-often anchored on a foundation that is skeptical of some the concepts discussed in this chapter. But with the mathematical and computational tools available now, there is little reason for that anymore.[11] What are some of the implications of complex systems theory to our goal of elucidating brain functions and how they relate to parts of the brain?

Interactions between parts   The brain is a system of interacting parts. At a local level—say, within a specific region—populations of neurons interact. But interactions are not only local to the area, and a given behavior relies on communication between many regions. Anatomical connectivity provides the substrate for interactions that span multiple parts of the cortex, as well as bridging the cortex, subcortex, midbrain, and hindbrain. This view stands in sharp contrast to a "localizationist" framework that treats regions as relatively independent units.

Levels of analysis   This concept is related to the previous one but emphasizes a different point. All physical systems can be studied as multiple levels, from quarks up to the object of interest. Not in all cases is it valuable to study the multiple levels (worrying about quarks in aerodynamics, say). But in the brain, studying multiple levels and understanding their combined properties is *essential*. One can think of neuronal circuits from the scale of a few neurons in a rather delimited area of space to larger collections across broader spatial extents. Multiple spatial scales will be of interest, including large-scale circuits with multiple regions spanning all parts of the nervous system. A possible analogy is the investigation of the ecology of the most biodiverse places on earth, including the Amazon rainforest and the Australian Great Barrier Reef. One can study these systems at very different spatial scales, from local patches of the forest and a few species to the entire coral reef with all its species.

Time, process   Complex systems, like the brain, are not static—they are inherently dynamic. As in predator-prey systems, it is useful to shift one's perspective from one of simple cause-and-effect to that of a *process* that evolves in time—a natural shift in stance given the interdependence of the parts involved. When we say a "process," there need not be anything nebulous about it. For example, in the case of three-body celestial orbits under the influence of Newtonian gravity, the equations can be precisely defined and solved numerically to reveal the rich pattern of paths traversed (Šuvakov and Dmitrašinović 2013).[12]

Decentralization, heterarchy   Investigating systems in terms of the interactions between their parts fosters a way of thinking that favors decentralized organization. It is the coordination between the multiple parts that leads to the behaviors of interest, not a master "controller" that dictates the function of the system. In many "sophisticated" systems, and the brain is no exception, it is instinctive to think that many of its important functions depend on centralized processes. For example, the prefrontal cortex may be viewed as a convergence sector for multiple types of information, allowing it to control behavior (see chapter 7). A contrasting view favors *distributed* processing through interactions of multiple parts. Accordingly, instead of information flowing hierarchically to an "apex region" where all the pieces are integrated, information flows in multiple directions without a strict hierarchy. An organization of this sort is termed a *heterarchy* to emphasize the multidirectional flow of information.

Emergence   Emergent properties are the norm in a complex system such as the brain. Of course, this does not invite fuzzy explanations. Instead, descriptions must be sufficiently detailed to allow *system-wide* properties to be captured.

Complexity   The property of complexity is a reminder that systems behave in ways that are substantially more varied and nuanced than at first entertained. Small changes of input or perturbations to their state can lead to *qualitatively* different behaviors and outcomes. This doesn't mean that complex systems are highly.unstable and erratic. In fact, complex systems live somewhere between complete predictability and total randomness.

9 500 Million Years of Evolution

What does evolution tell us about the organization of the brain when we consider all vertebrates groups: mammals, birds, reptiles, amphibians, and fishes? To appreciate what evolution teaches us, we need to keep track of several moving pieces so that we can compare neuroanatomies across taxonomic groups—say, fishes and birds. Although this makes for potentially more challenging reading, we'll learn important lessons as we identify similar structures (such as the striatum) and connectional systems (such as the one involving the basal ganglia) across the vertebrates. Unsurprisingly, there are many important differences, too, even involving structures that neuroscientists like to refer as "conserved" (like the amygdala). Together, chapters 9 through 11 make up a unit that puts the different parts of the brain together into an interconnected whole, starting from an evolutionary perspective that helps discern a broader picture of brain organization.

When we think of a brain, most of us conjure images of the human cerebral cortex—the outer surface with its protrusions and grooves. But how does it vary across animals? The brain of a human weighs more than 1,000 grams, that of a rhesus monkey around 100 grams, and that of a marmoset less than 10 grams. Despite spanning two orders of magnitude in weight, they are quite similar. But how about the brain of a dolphin or an elephant (a large African elephant's brain weighs over 5,000 grams)? If the similarity of the first three species (they are all from the primate order) wasn't curious enough, the resemblance of all five will be a surprise if you haven't seen open specimens in a science fair or a zoo. The species discussed thus far are all mammals. But how about the brain of a salmon, a common frog, an alligator, or a crow?

We are now talking about vertebrates, which together cover a very broad range of body types and life styles, from aquatic to terrestrial to aerial. The evolutionary timescale here is truly mind-bending. The common ancestor to all vertebrates inhabited earth more than 500 million years ago (figure 9.1).

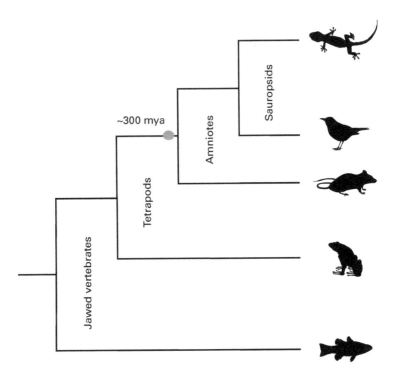

Figure 9.1
Evolutionary tree of living vertebrates. Note that the evolutionary trajectories of mammals and sauropsids (reptiles and birds) diverged over 300 million years ago (mya).

Amniotes are animals that have an amniotic sac that provides nutrition to the embryo; they include reptiles, birds, and mammals. Note that the evolutionary trajectory of mammals diverged from that of reptiles and birds (sauropsids) more than 300 million years ago. This underappreciated fact means that mammals are not descendants of reptiles and, as such, there's no reptilian brain inside the mammalian brain! At first glance, the brain of a fish looks rather different from that of a human (figure 9.2). But what do we find if we dissect it and carefully analyze it? Anatomists started doing just that at the end of the nineteenth century, a time when evolution was establishing itself as a principle at the core of biology.

Unfortunately, evolutionary thinking would be colored with the notion of "progress toward advanced forms" well into the second half of the twentieth century, and descriptions of evolution in terms of an ascending ladder with humans at the top were customary. T. H. Huxley, the academic who

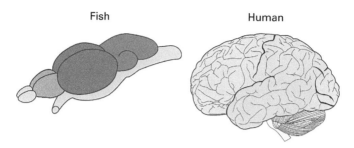

Figure 9.2
Considerable macroscopic differences exist between the fish and human brain.

fought many intellectual battles in defense of Darwin, may have been the first to describe brain evolution as proceeding linearly from fish to human (Striedter 2005). Indeed, Victorian England scientists were prompt to reuse the old Aristotelian concept of the *scala naturae*: elements can be ordered into a series of steps ascending from rocks all the way to the almighty.

Although this view is no longer held by scientists, it is at times hard to break from thinking of evolution in terms of "directed progress," which places mammals, if not primates or the great apes, at some higher level and everyone else below. But properly understanding the evolutionary history of vertebrates is critical for thinking about the human brain, not least because of the popular view that a "reptilian brain" lies deep inside the primate brain. This notion, repeatedly incessantly by nonbiologist researchers and scholars (and still a few neuroscientists!), is glaringly erroneous and severely distorts what the study of the anatomy across vertebrates—comparative neuroanatomy—has taught us in the past decades.

The Basic Chassis

The evolutionary path of the vertebrates is a story of over half a billion years. The central nervous system of all vertebrates has three major components: the hindbrain, the midbrain, and the forebrain. Recall that the hindbrain includes the lower part of the brainstem; the midbrain includes the upmost part of the brainstem; and the forebrain includes subcortical structures like the thalamus and, in mammals, all of the cortex. Although brains differ considerably in overall shape across classes (mammals, reptiles, and so on), and indeed within a class, they all contain these macroscopic components. In a nutshell, vertebrate brains have a three-part chassis. But as we will see, the commonalities extend well beyond this coarse level of organization (figure 9.3).

In mammals, the vertebrate forebrain includes the cortex dorsally and the subcortex ventrally. In other vertebrates, the forebrain also has dorsal and ventral parts, but because they are structurally different from that of mammals, anatomists call them "pallium" (meaning cloak in Latin) for the dorsal part and "subpallium" for the ventral part (every time you read these terms, go back to figure 9.3 to remind yourself). So, in this chapter we will use these two terms when talking about the forebrain. In mammals, the pallium contains all of the different types of cortex, from the hippocampus with a basic three-layer organization to the so-called isocortex with six layers. In nonmammals, the pallium isn't organized in a layered fashion but instead has clusters of cells with different properties and arrangements.[1] By following brain development, neuroanatomists can identify subsectors of the pallium that are common to all vertebrates. They are unimaginatively labeled "dorsal," "ventral," "medial," and "lateral." Thus, when trying to understand how the brain of, say, a reptile and a mammal map to each other, it is important to keep track of these sectors, as we'll discuss later.

The mammalian subpallium is comprised of cell masses at the base of the forebrain, including the striatum and parts of the amygdala, both of which are found in all vertebrates. But several other regions are found in common, too, including the hypothalamus and the thalamus. In the roof of the midbrain, we find the optic tectum (called superior colliculus in mammals), which we studied in chapter 3.

We see, therefore, correspondences at three levels at least: that of broad territories, such as the forebrain; that of their subdivisions, such as pallium versus subpallium; and that of areas like the amygdala. Working out such correspondences is one of the central goals of evolutionary neuroscience. The challenge is one of establishing mappings between subparts that can be quite distinct; the brain of a dog is not an enlarged version of the brain of a salmon. The problem is fiendishly complex, as it's not clear what criteria should be applied. For example, the amygdala of a mammal is not just there for the picking in a bird or a reptile. So, is there one? And if so, how should it be defined? Should we recognize the amygdala in, say, a fish based on its position in the forebrain, cellular composition, gene expression pattern, anatomical pathways, and function? And how should these factors

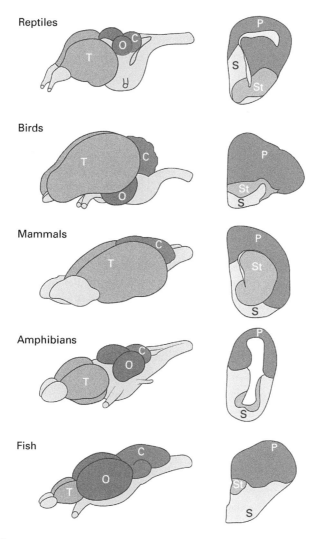

Figure 9.3

Vertebrate brains. All vertebrate brains contain corresponding sectors, including the telencephalon (T), the optic tectum (O), and the cerebellum (C). The optic tectum, which is called superior colliculus in mammals, is not visible from the outside in mammals and lies at the top of the midbrain, next to the periaqueductal gray (see figure 5.1). The leftmost part in all brains is the olfactory lobe. On the right, cuts through the brain are shown at the level of the telencephalon, revealing the pallium (P), subpallium (S), and striatum (St), which are present in all vertebrates.

be weighted? What's more, these variables are not static but change during embryological and developmental stages, from the earliest phases of the central nervous system to the adultlike, completed form. Not surprisingly, comparative neuroanatomy remains a contentious field of research, and arguments can get very heated.

No Rat Was Ever an Ancestor of Any Monkey

Not a single fossil brain has ever been recovered.[2] Fossilization is possible for hard body parts, such as bones and teeth, and the brain definitely isn't one of these. With roughly the consistency of jello, it needs to be encased in an armored cranium to be kept safe. Absent fossil brains, how can we use evolution to understand changes to the brain? Broadly, the strategy has two main components. First step: Select animals that are descendants of a common evolutionary lineage. If you are interested in the human brain, you could study the brain of *living* monkeys and rodents, or you could extend the timeline and consider all tetrapods (that is, amphibians, reptiles, birds, and mammals). Second step: Assume that the brain of *current*, living animals in question are sufficiently similar to the *ancestral* forms, so that some inferences about the ancestral forms can be drawn.

Take the first step. The extensive fossil record and careful examination of body parts provides a foundation for understanding the evolution of major lineages. Add to that genetic information gathered in the past decades and a good picture emerges. Now to the trickier second step. Here, we must recognize that animals chosen to represent ancestral groups are not the actual ancestors—they are descendants. They are suitable for phylogenetic comparisons because they possess many features that are *primitive*—that is, relatively unchanged from ancestral forms. The primitive feature of a group can often be *inferred* by comparing those of the living members of the group and looking for elements *common to all* (such as four major appendages or limbs in vertebrates). However, the greater the degree of diversity and specialization within a group (such as fins in fish and forelimbs in reptiles), the greater the need for studying more variants. In the case of living animals, the assumption is usually made that the more primitive characteristics a given species has, the more likely it is to resemble the ancestral form. The upshot is that step two will always involve a considerable, though acknowledged, uncertainty. As stated by a prominent researcher of brain evolution, this field of study is not for those averse to uncertainty! Nevertheless, an evolutionary perspective is absolutely required for a deeper grasp of the structure of the human brain—and, in fact, all brains.

By now the title of this section should be clearer to you. A rat, a living species, cannot be an ancestor to a human, another living species (or vice versa). In fact, 75 million years separate humans and rodents. That is to say, their common ancestor diverged approximately 75 million years ago, a window of time during which the brains of both lineages have evolved.

Consider figure 9.3 once again. The telencephalon is identified in all major vertebrate taxonomic groups. The conclusion drawn is that this brain sector was present in the common ancestor to all vertebrates, which was trekking around 500 million years ago. In one sense, this is a gigantic inferential step. So why is it made? One can gain a handle on the question by considering alternative explanations. The central one is called "convergent evolution," the process by which the same solution (here, creating a telencephalon) is independently reached and *not* inherited from a common ancestor. In the present case, perhaps the telencephalon was created from scratch for each of the major vertebrate groups. This alternative, though logically possible, is considerably less parsimonious than the one based on common ancestry. But perhaps it could happen, right? No, once we consider the large amount of data about the telencephalon, including what we know about embryology, genetics, cell types, anatomical connectivity, function, and more, the second solution starts looking absurd—the odds of such scenario would be vanishingly small.

Changing Views on Animal Cognition

For most of the twentieth century, scientists had vertebrates other than mammals in low esteem. These were creatures with a narrow stock of stereotypical behaviors, largely stimulus driven (give it an input, and a more or less fixed action ensues), and confined to the here and now.

In recent decades, our view of animals' behavioral capabilities has witnessed a sea change thanks to striking discoveries from ethologists (who study behavior as it occurs in natural environments) and comparative psychologists (who study behavior across species). Field studies have revealed complex behaviors across phylogenetically distant taxonomic groups in vertebrates (and even invertebrates). New approaches and techniques in the laboratory have been employed to elucidate the mechanisms underlying behaviors in different species. The emerging picture is one in which behavioral plasticity—the extent to which they can be modified—and flexibility are widespread in the animal kingdom.

Take, for example, spatial cognition: how animals navigate through the world and process positional information. All vertebrates have to move efficiently within their environment and thus need to learn and retrieve the location of different resources or threats. Different species of mammals, birds, reptiles, amphibians, and fishes exhibit parallel spatial abilities, such as those involved in homing behavior, spatial navigation, and spatial learning. Another research area of growing interest is social behavior. The ability to learn from others, discriminate among individuals, and categorize them as offspring, mates, rivals, allies, or neighbors is common to most vertebrate classes.

Perhaps nowhere are sophisticated behaviors more remarkable than in some birds (Emery and Clayton 2004). A person who has wronged a crow in the past is promptly recognized, scolded, and mobbed. Careful lab studies show that corvids (including crows) and parrots solve problems and have complex episodic memory capabilities. The former involves both tool use and manufacture, and the latter includes "mental travel" in time and space, such as retrieving information about the "what, where, and when" of experiences. Indeed, the intelligence of some birds rivals that of the great apes and dolphins, if not actually surpassing it.

Amphibians and reptiles show many traits common to those found in birds and mammals, including elaborate forms of communication, problem solving, parental care, play, and complex sociality.[3] Fishes learn spatial tasks and engage in social interactions driven by repeat interactions with the same clients (such as other fish species that they clean). Looming evidence even indicates that fishes engage in problem solving and invent tools; for instance, wrasses use rocks as anvils to crack clam shells. And to think that not long ago, the use of tools was believed to be an exclusively human capability!

It is now abundantly clear that vertebrates other than "advanced" mammals are a far cry from being rudimentary automatons. Unraveling how these behavioral capabilities are enabled by the brain is the challenge now.

Evolving Our View of Brain Evolution

In 1896, the German anatomist Ludwig Edinger published *The Anatomy of the Central Nervous System of Man and Other Vertebrates*. The book, which established Edinger's reputation as the founder of comparative neuroanatomy, described the evolution of the forebrain as a *sequence of additions*, each of which establishing new brain parts that introduced new functions.

Edinger viewed the forebrain as containing an "old encephalon" found in all vertebrates. On top of the old encephalon, there was the "new encephalon," a sector only more prominent in mammals. In one of the most memorable passages of his treatise, Edinger illustrates his concept by asking the reader to imagine precisely inserting a reptilian brain into that of a marsupial (a "simple" mammal). When he superimposed them, the difference between the two was his new encephalon. He then ventures that, in the brain of the cat, the old encephalon "persists unchanged underneath the very important" new encephalon (Edinger 1908, 446). Put differently, the part that was present before is left unaltered. Based on his coarse analysis of morphological features, Edinger's suggestion was reasonable. But to a substantial degree, his ideas were very much in line with the notion of brain evolution as progress toward the human brain—à la old Aristotle and the *scala naturae*. Given the comprehensive scope of Edinger's analysis across vertebrates, his views had a lasting impact and shaped the course of research for the subsequent decades.

More than a century later, knowledge about the brains of vertebrates has expanded by leaps and bounds. Yet, old thinking dies hard. Antiquated views of brain evolution continue to influence, if only implicitly, neuroscience. As an example, bear in mind that most frameworks of brain organization are heavily centered on the cortex. These descriptions view "newer" cortex as controlling non-cortical regions, which are assumed to be (relatively) unchanged throughout eons of evolution. Modern research on brain anatomy from a comparative viewpoint indicates, in contrast, that brain evolution is better understood in terms of the *reorganization of large-scale connectional systems*. We will develop this idea in the remainder of the chapter having in mind particular parts of the brain. To set the stage, figure 9.4 illustrates the overall organization of connectivity in vertebrates.

Decoupling Sensory Signals from Motor Responses

In chapter 3, we discussed a circuit involved in both defensive and appetitive behaviors centered on the optic tectum/superior colliculus of the midbrain. This system is extremely important across vertebrates. In rodents,

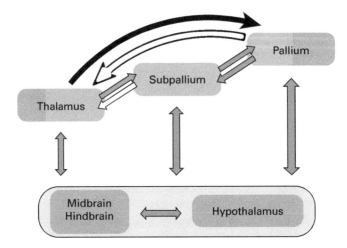

Figure 9.4
Overall anatomical connectivity plan of the vertebrate brain. Pathways shown by white arrows aren't present in fishes or amphibians. The black pathway from the thalamus to the pallium is very weak in fishes, more pronounced in amphibians, and considerably more substantial in reptiles, birds, and mammals.

it helps the animal decide if it should flee when movement is detected overhead or possibly approach and explore further if the movement is in the lower visual field. But the animal's behavior is flexible and not fixed by the input—the context in which it occurs, encompassing both external and internal worlds, is critical.

But how about toads and frogs (the Anuran family), with more limited behavioral repertoires?[4] The optic tectum allows Anurans to tongue-snap when stimuli are in certain parts of their visual field, like an insect flying overhead, a reaction considered to be rather automatic. In these animals, the optic tectum is described as a sensorimotor interface because it receives retinal projections carrying visual signals and projects to brainstem and medullary motor circuits (see chapter 3), being thus well positioned to eject the tongue when attempting to capture prey.

Both motivational and attentional factors mold these behaviors (figure 9.5). During the mating season, prey-catching is minimal, and other behaviors are favored (not surprisingly, males approaching females). Regarding attention, when an animal is prepared to attack, the presence of prey causes animals to reorient themselves in a way that favors the strike; physiologically,

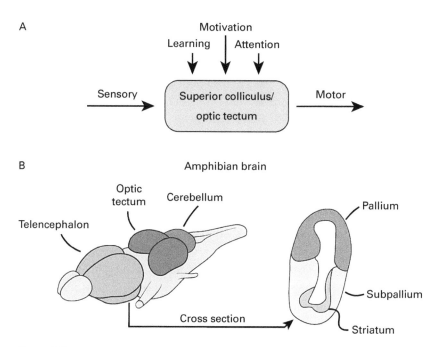

Figure 9.5

Decoupling inputs from outputs. (a) Input stimuli can trigger actions very directly through the superior colliculus/optic tectum. Nevertheless, the animal's motivational and attentional state, as well as past learning, contextualize the responses, bringing them relative flexibility. (b) Amphibian brain highlights some of the parts discussed in the text.

the related sensory cues evoke considerably stronger cell firing in the optic tectum. Prey-catching is also tuned by learning. In one study, toads that were hand-fed on mealworms later responded to the hand alone, demonstrating an association between food and the initially meaningless stimulus (the hand). Together, these examples show how an animal's responses, possibly to the same input, are shaped by diverse variables.

Back in chapter 3, we saw how the optic tectum's participation in multiple circuits allows it to generate context-dependent actions. For one, signals in the hypothalamus reflect several motivational variables, including sex-related ones that fluctuate seasonally. Hypothalamic outputs can influence the optic tectum through projections to its deep layers, which are the ones that have direct connections to motor areas.[5] Other long-range circuits play a role in the case of learning. The circuit involving the medial pallium is necessary for both learning associations and for using the acquired information later.

Intriguingly, the medial pallium in mammals is the sector of the forebrain that forms into the hippocampus, a simple type of cortex with only three loosely organized cell layers. In chapter 11, we will discuss the hippocampus in more detail, including its roles in memory and spatial navigation.

No doubt, context sensitivity and behavioral flexibility are more limited in amphibians compared to reptiles, birds, and mammals, which produce a broader range of behaviors. So, let's consider the architecture of the optic tectum in other vertebrates, with the anatomical connectional systems that interlink disparate brain parts.[6] In reptiles, the basic connectional architecture is noticeably enlarged. The thalamus is more richly connected with other parts of the forebrain and connects to additional sectors of the pallium. In birds and mammals, the connections of the optic tectum with the thalamus and pallium are, in turn, extensively developed compared to those in reptiles. Overall, the number of potential long-range circuits is quite large. In mammals, in particular, the overall connectivity of the superior colliculus is enormous.

Why such a degree of complexity involving long-range circuits that span the midbrain, thalamus, and pallium/cortex (see figure 9.4), even in "simple" animals (if one applies this label to most vertebrates including small mammals)? I suggest that it confers a high degree of behavioral flexibility allowing animals to cope with the multifaceted interactions they engage in involving predators, prey, potential mates, and so on. In species with more malleable behaviors, survival benefits from circuits that can form *combinatorially*—from region A to region B via multiple routes—as the number of conditions related to the internal and externals worlds of the animal are exceedingly high. We will return to this idea in chapter 10.

Returning to a principle outlined in chapter 3, another way to think about this type of organization is that it *decouples* sensory and motor elements: Sensory signals do not necessarily trigger motor actions; when an action ensues, the sensorimotor transformation takes into account an array of influences, and sensation and action are part of a continuous loop that can flexibly update itself (that is, acting on the world changes the information that is sensed, leading to revised actions). Multiple variables are entertained that cancel, enhance, or otherwise refine the types of actions undertaken. As we'll see below, this decoupling property is not particular to the optic tectum but is an essential element of the vertebrate brain—from "simple" to "sophisticated" animals.

The Great Loops of the Basal Ganglia

The striatum and adjacent regions at the base of the brain are collectively referred to as the basal ganglia (chapter 2). Neurodegenerative diseases that affect this system include Parkinson's disease and Huntington's chorea. Parkinson's is related to cell loss in the dopaminergic regions of the midbrain that project to the striatum and compromises the ability to initiate voluntary movements. Huntington's impacts the major output projection neurons of the basal ganglia and is characterized by uncontrolled movements.

A major discovery in the 1970 and 1980s was that in mammals, the basal ganglia work in close coordination with the cortex through a loop-like circuit. For example, motor and somatosensory cortex project to the striatum, which connects back to the cortex through the thalamus (figure 9.6). Strikingly, the basal ganglia are involved not only in this movement-related circuit, but are part of multiple loops. Whereas sensorimotor cortical areas target dorsal parts of the striatum, other parts of the cortex project to more ventral ones, including the nucleus accumbens.[7] Again, the circuit loops back to the cortex through the thalamus. Given the participation of the accumbens in motivational processes and its connections with regions such as the amygdala, this circuit is frequently labeled "limbic," but as we saw in chapter 6, this term is next to meaningless.

The pathways interlinking the cortex with the basal ganglia reveal that the two work in a coordinated fashion. Given the prominence of this arrangement in mammals, are there comparable features in other vertebrates? Not only are the subregions that make up the basal ganglia present across vertebrates, but loop-like circuits are found in the tetrapods (amphibians, reptiles, birds, and mammals). This is remarkable as it shows that an elaborate circuit was most likely a property of a common ancestor to all tetrapods. But there are notable differences, too. Amphibians and reptiles only have loops involving the *ventral* parts of the striatum. Birds have circuits coursing through both *ventral* and *dorsal* striatal territories, just like in mammals. Importantly, the connectivity in both birds and mammals is substantially more developed, at once more extensive and with stronger pathways.

What are some of the implications of the basal ganglia template across tetrapods? To answer this question, we need to consider the organization of the pallium in vertebrates and which of its sectors project to the ventral striatum. Ventral striatum circuits have a major influence on the energy and vigor of behavioral responses—the amount of *effort* that is exerted by

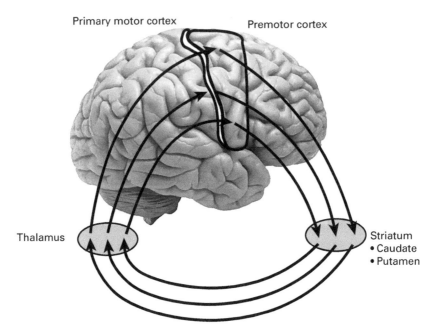

Figure 9.6
Cortical-subcortical loops. The cortex and subcortex work in close coordination through systems of pathways forming loops. The primary motor cortex is indicated by the thick white stripe.

an animal. Lesions that compromise the ventral striatum impair how much effort is exerted in the pursuit and avoidance of rewarding and punishing stimuli. In particular, the ventral striatum helps *invigorate* the animal to approach a reinforcing stimulus—that is, a stimulus that has the potential to lead to reward.[8] In nature, prey, mates, and so on are never at arm's length. It's a basic fact of life that performing work and tolerating delays (waiting) are necessary for attaining positive outcomes.

It is only possible to obtain motivationally relevant items by engaging in behavior that brings them closer or makes their occurrence more likely. In fact, because animals are usually separated from reinforcing items by a long distance or by various obstacles, effective behaviors require work, such as foraging for food. Animals must thus allocate considerable resources toward "seeking behaviors," which vary in terms of speed, persistence, and overall level of "output." Although the exertion of effort can be relatively brief at times (for example, a predator pouncing on a prey), under many circumstances it must be sustained over long periods of time. Thus, effort-related capabilities are highly adaptive—they are advantageous for survival—because in nature survival depends on how well animals over-come work- and time-related response costs.

Let's return to basal ganglia loops. Ventral basal ganglia loops connect the pallium with the ventral striatum, which, as discussed, is a major *moti-vational hub* that helps regulate the amount of vigor and energy expended. We find an interesting property across tetrapods. In amphibians, two major sectors of the pallium have loops with the ventral basal ganglia. In reptiles and birds, this number grows to three, and in mammals four. It stands to reason that the number of pallial sectors that engage in loops with the basal ganglia—as few as two and as many as four—determine the types of signal from the pallium that have a more direct impact on actions. The larger the number of sectors, the more diverse the signals from the pallium that impinge on the striatum, allowing a broader range of variables and their combinations to influence behaviors.

But the possibilities in birds and mammals are enlarged further (figure 9.7). They have an expanded set of basal ganglia loops, as circuits course through the dorsal striatum in addition to the ventral striatum. And, again, these loops involve very diverse parts of the pallium; in both birds and mammals, all four pallial sectors project to the dorsal striatum. The dorsal parts of the basal ganglia process sensory and motor signals, which are distinctly impor-tant in the control of finer movements—like reaching out to grasp a ripe fruit. Furthermore, as behaviors develop temporally as one action merges into another, basal ganglia loops allow the continuous adjustment of move-ment parameters and goals ("move here; now veer to the left some"). That is to say, the basal ganglia support the generation of actions in a dynamic fashion.[9] As motor behaviors unfold, cortical-basal ganglia loops continu-ously update motor programs so as to reflect the most recent data.

More so than in other vertebrates, in birds and mammals, processing in the pallium spans a wide range of abstractness and, in many instances, is not closely tied to sensory and motor variables. Thus, loops involving more regions of the pallium with "abstract" properties support complex spatiotemporal behaviors, including exploiting available goods, exploring the environment, and avoiding threats. By bringing the pallium to bear on actions, they provide a scaffold for more flexible and sophisticated behav-iors. In a very real sense, the extensive projections from all major sectors of

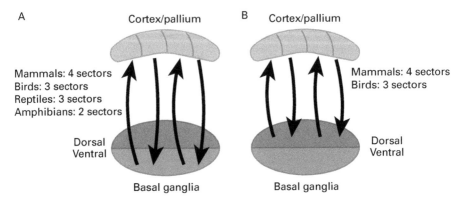

Figure 9.7
Cortex-subcortex loops via the basal ganglia across vertebrates. (a) Ventral loops are present in all vertebrates, except fish, but vary in the number of sectors of the cortex/pallium they involve. (b) Dorsal loops are only observed in mammals and birds.

the pallium to the striatum in birds and mammals bring the multitude of functions of the pallium/cortex into the circuits of the striatum.[10]

Thus far, we have not discussed the most numerous and diverse vertebrates on the planet. How are the basal ganglia organized in fishes? As in other vertebrates, the pallium projects to the striatum. But here a rather different type of arrangement is found: The striatum projects directly to the pallium, which is a feature not seen in other vertebrates where the circuit is always indirect and courses back to the pallium by way of the thalamus. What is especially interesting is that the striatum of fishes is a convergence area where pathways from many parts of the brain impinge, including the hypothalamus, midbrain, and hindbrain, in addition to thalamus and pallium. In all likelihood, pallium-striatum circuits communicate and exchange a broad spectrum of signals. As fish don't make the most conducive lab animals, little is known about their neurophysiology. But it would be revealing to investigate diverse species because their telencephalon is quite variable. In some cases, it is rather pronounced, and as mentioned, some fish display sophisticated behaviors, even making simple tools.

Large-Scale Circuits of the Amygdala

It is common for neuroscientists to think of the amygdala as triggering immediate, obligatory emotional responses. We discussed in chapter 5 the two major amygdala sectors: basolateral and central. In mammals, the pallium includes all forms of cortex. Thus, it will be surprising that the basolateral amygdala is part of the pallium: Embryologically, the tissue that eventually forms the pallium produces this sector, too. In contrast, the central amygdala is part of the subpallium. So, these two parts of the amygdala, which interact so strongly, are actually structurally rather distinct creatures, more like distant cousins than siblings. The different origins of the two amygdala components also helps explain their different connectivity profiles, as further discussed below.

Given the reputation of the amygdala as a "primitive" structure, one might imagine that identifying it across vertebrates would be straightforward.[11] This is far from true. For one, the amygdala straddles the pallium and subpallium, and their boundary is very challenging to track with confidence. Additionally, the deep evolutionary split between mammals and sauropsids (reptiles and birds), separated as they are by about 300 million years, has thus far prevented comparative neuroanatomists from conclusively determining the correspondence between some regions and even larger parts of their brains. In sauropsids, for example, a rather prominent part of the forebrain (known by the unhelpful name of "dorsal ventricular ridge") is not easily mapped to mammalian features. As it turns out, the basolateral amygdala is at the center of current scientific debates (and battles!) on how to understand the brain of mammals and sauropsids.

In chapter 5, we considered how, in mammals, the anatomical pathways of the basolateral and central amygdala are strikingly different. The basolateral amygdala is interlinked bidirectionally with most of the cortex, from occipital to frontal. In particular, the basolateral amygdala is a convergence site for all sorts of sensory information, thus in a privileged position to simultaneously take into account the environment and the body, helping the animal segregate the significant from the mundane. The central amygdala, instead, has extensive interconnections with the hypothalamus and brainstem nuclei (including sites in the midbrain, pons, and medulla) involved in behavioral, autonomic, and neuroendocrine responses (there are some pathways between the central amygdala and cortex, too). If we consider that the basolateral amygdala is pallial in origin but the central amygdala originates in the subpallium, it is perhaps a little less mysterious how these two regions of the brain, close enough to be joined together under the same umbrella—"the amygdala"—associate anatomically with such different partners.

In birds, a pallial amygdala-like region has been identified, too, and exhibits extensive interconnectivity with other regions of the telencephalon in a way that strongly resembles the mammalian counterpart. For example, it receives multiple sensory inputs from other pallial areas as well as inputs from "association" areas in the pallium that are not so directly linked to sensory processing. (Neuroscientists call areas of the pallium/cortex "associational" because they combine multiple perceptual attributes.) Because these connections are bidirectional, the pallial amygdala can influence these regions, too. In reptiles, as in birds, a pallial amygdala-like region also has a broad range of telencephalic pathways, likewise interconnecting it with sensory and association areas of the pallium. The pallial amygdala is actually a very prominent association center of the reptilian brain—a region that influences and is influenced by multiple areas of the pallium.

Overall, in amniotes (mammals, birds, and reptiles), the pallial amygdala is a major hub region of the telencephalon. But the interconnectivity doesn't stop there. This sector also projects to the subpallium, including the central amygdala and the hypothalamus. Of particular interest here are the connections of the pallial amygdala with both the dorsal and ventral basal ganglia. In the previous section, we discussed the arrangement of dorsal and ventral basal ganglia loops and their roles in shaping vertebrate behaviors. The pathways from the pallial amygdala to the striatum intertwine the connectivity of the former with the loops of the latter, creating a giant network of connectivity (figure 9.8).[12] Taken together, the pallial amygdala of birds and mammals, in particular, is in a focal position to integrate disparate information and to influence both emotional and motivated behaviors, in addition to cognitive functions. Bearing in mind that the connectivity is more restricted (both weaker and connected to fewer places) in reptiles, the pallial amygdala likely plays a comparable role in these animals, too.

How about the case of fishes and amphibians? A pallial amygdala-like area has been identified in amphibians. In fishes, stronger evidence of a related area has been obtained for teleost fishes. (Teleost fishes comprise up to 96 percent of existing fish species and have modifications of the musculature that allows them to protrude the jaw outward from the mouth; this is behaviorally advantageous because it allows them to grab prey and draw them into the mouth.) In teleost fishes and amphibians, the pathways of the pallial amygdala are reminiscent of the amniote organization, including connectivity with other pallial regions, although these connections are scarcer.

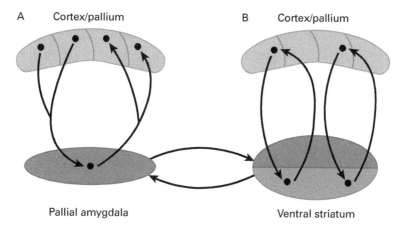

Figure 9.8

Interlinking of circuits. The cortical-subcortical circuits of the pallial amygdala are interconnected with ventral basal ganglia loops.

We started this section saying that neuroscientists like to describe the amygdala as producing basic, obligatory emotional responses, such as generating an alarm call when a predator is spotted by a prey. True, this is an important function of the amygdala—or to be precise, it is an important function of one of the two amygdalas, the central one, which is part of the subpallium. But this grossly oversimplifies—in fact, mischaracterizes—the complexity and scope of the interactions that the pallial amygdala of all vertebrates engages in. In the end, the contributions of the amygdala involve all major dimensions—sensory, motor, emotional, motivational, and cognitive.

Conservation of Structure and Function

The amygdala of primates can be identified in other mammals—that much is easy. Researchers thus invoke the idea of evolutionary *conservation* to stress that the amygdala is present among evolutionary lines stretching tens of millions of years. For example, primates and rodents diverged more than 75 million years ago (before the extinction of the non-avian dinosaurs)—their common ancestor roamed the planet around that time.

On the surface, conservation appears to be a straightforward concept: Identify a feature (cortex, region, or molecule) in an animal line which is also present in a distantly related line and, voilà, it is established. Yet the situation is far from simple. As can be gathered from the previous section, a brain region doesn't come in exact copies to be found in an amphibian and a mammal, for instance. Is the amygdala in an amphibian conserved if it has only a central and not a basolateral part as in a mammal? How about if both sectors are found but their anatomical connectivity is substantially different? Or they have substantially different cell types? In other words, what should count to establish conservation?

We need here the concept of *homology*. In evolutionary biology, two structures are homologous if they were present in a common ancestor. Thus, the cortex is thought to be homologous across mammals because it is present in the ancestor of all mammals. But homology doesn't depend on *function* at all; the common origin is the one that matters. For example, fins and hands are homologous even though they serve entirely different purposes. Still, conclusive genetic evidence shows that they are derived from genetic regulatory systems present in the common ancestor to the vertebrates.

But if homology is independent of function, what is exactly conserved? Only genetic programs? Indeed, the question is baffling enough that evolutionary biologists have been pondering it for a long time.[13] At its core, the question confronts the seemingly impossible problem of elucidating what is *new* in biology. If animal species derive from common ancestors by a process of descent with modification—the central tenet of Darwinian thinking—at what point is a feature (also called a character) of the phenotype truly novel? Needless to say, I will not be foolish enough to attempt to answer this question, but I suspect that, like in many other cases, part of the problem lies in phrasing it as a dichotomy in the first place—novel or not?

Armed with the above ideas, let's revisit the evolution of the basal ganglia. The basal ganglia are deeply conserved because essential components are identified across tetrapods. Does this mean that the basal ganglia of a marmoset monkey are roughly the same as that of a toad? We know this can't be true because the amphibian basal ganglia only have loops via the ventral striatum, not the dorsal; the mammalian basal ganglia have loops involving more territories of the pallium than the amphibian one. And, of course, such changes will have functional ramifications. Perhaps signals will be combined in mammals in ways that are not possible in amphibians or, alternatively, be more segregated from one another. That is to say, the new components present in mammals will alter the circuit's computational capabilities. Thus, brain evolution involves changes to pathways that have the potential to bring about significant functional modifications.

This view is quite different, of course, from seeing evolution as adding new brain parts atop older ones. Let's develop the idea a bit further by thinking about the appearance of the cortex in mammals. As stated, according to older views, "newer" cortex controls subcortical regions, which are assumed as a rule to be relatively unchanged throughout evolution. The enlargement of the forebrain with the addition of multilayered cortex is seen through the lens of hierarchical organization. In contrast, if both the cortex and subcortex change, as proposed here, they may change in a coordinated fashion—in the resulting circuitry, the cortex and subcortex are mutually *embedded*.

The amygdala provides a good example of this type of joint embedding. One study found that parts of the basolateral amygdala are considerably more "developed" in monkeys than in rats (Chareyron et al. 2011). The authors suggested that the differences in the relative subregion size and neuron numbers between the two species are related to the connectivity of this sector. In this manner, the "enhanced" properties of the monkey basolateral amygdala parallel the greater development of the cortical areas with which the basolateral sector is interlinked. Such correlated evolution likely supports higher convergence and integration of information in the basolateral amygdala. Irrespective of the mechanisms behind these evolutionary changes, differences between species are considerable. Studies comparing humans, apes (such as chimpanzees and gorillas), and monkeys discovered that the number of neurons in parts of the basolateral amygdala are 50 greater in humans.[14] Such substantial differences are rarely seen in comparative analyses of human brain evolution. For example, the volume of the human cortex is 24 percent larger than expected for a primate of our brain size, whereas the human frontal lobe, frequently assumed to be enlarged, is approximately the size expected for an ape of human brain size.

Understanding evolutionary conservation is far from a technical issue or armchair musing. The National Institutes of Health in the United States and funding agencies around the world invest billions of dollars in brain research in the hopes of curing, or at least ameliorating, conditions that stem from brain malfunction. Simply put, most of the research cannot be done ethically in humans, and animal models of diseases are, at present, the only way forward. The assumption, of course, is that by studying the mouse or rat brain, for example, we will gain important knowledge that is transferable to deciphering the human system. But essential components of brain function have to be conserved for this approach to be sound. Our brief incursion into the evolution of the vertebrate brain in this chapter shows that, while studying animal models is certainly informative, we need to proceed with caution. Only by studying a broader range of animals will it be possible to clarify how varying neural architectures support behaviors. In this sense, the heavy emphasis on studying mice and rats is very shortsighted.

To conclude: The human brain, or even more generally the mammalian brain, is not a sophisticated cortical machine built atop old, inflexible brain territories that only support simple, stereotypical behaviors. The anatomical architecture of vertebrates supports signal communication across all major brain territories, including between the pallium and subpallium in the forebrain and between the forebrain, midbrain, and hindbrain. At the same time, pathways vary considerably across taxonomic groups. Whereas long-distance circuits are present in fishes and amphibians, they truly flourish in the amniotes, especially birds and mammals.

10   The Big Network: Putting Things Together

If the brain is a complex system of interconnected parts, how do they fit together? Here, we discuss several general principles of brain organization that help us understand the whole. The massive anatomical connectivity linking regions creates a highway backbone that allows signals to travel very efficiently across the brain. From these physical pathways, functional relationships emerge based on how distant regions coalesce into functional units. From this perspective, *networks*, not regions, are the meaningful units in the brain. But even these networks are elusive because they have complex properties in both space and time. Nevertheless, having a better grasp of their properties is essential to clarify how the nervous system generates behaviors.

Engineers think of systems in terms of inputs and outputs. In a steam engine, heat (input) applied to water produces steam, and the force generated pushes a piston back and forth inside a cylinder; the pushing force is transformed into rotational force (output) that can be used for other purposes. Reasoning in terms of input-output relationships became even more commonplace with the invention of computers and the concept of a software program. Thus, it is only natural to consider the brain in terms of the "inflow" and "outflow" of signals tied to sensory processing and motor acts. During sensory processing, energy of one kind or another (e.g., light or sound) is transduced, action potentials reach the cortex, and are further processed. During motor acts, activity from the cortex descends to the brainstem and spinal cord, eventually moving muscles. Information flows *in* for perception and flows *out* for action.

For all but the simplest of reflexes, decoupling sensory inputs from motor outputs is necessary to confer any behavioral flexibility. Most of the brain is, of course, interposed between input and output. But how is this "black box" organized? In chapter 7, we saw that the "sequential model" was revised based on newer anatomical knowledge. Parallel processing streams, feedback between stages, and other forms of communication were introduced. In chapter 8, we discussed ideas about complex systems, including some implications for studying an object as multidimensional as the brain. In chapter 9, we applied an evolutionary approach to examine bidirectional interactions between the cortex and subcortex in the cases of the basal ganglia and the amygdala.

Now, I will outline a framework for thinking about large-scale brain circuits that I've called *functionally integrated systems*.[1] Before doing so, I will outline five broad principles of organization, establishing concepts that undergird these functional circuits. To anticipate, some of the consequences of the principles are as follows: The brain's anatomical and functional architectures are highly nonmodular; signal distribution and integration are the norm, allowing the confluence of information related to perception, cognition, emotion, motivation, and action; and the functional architecture is composed of overlapping networks that are highly dynamic and context-sensitive.

Principle 1: Massive Combinatorial Anatomical Connectivity

Dissecting anatomical connections is incredibly painstaking work. Chemical substances are injected at a specific location and, as they diffuse along axons, traces of the molecules are detected elsewhere. After diffusion stabilizes (in some cases, it takes weeks), tissue is sectioned in razor-thin slices that are further treated chemically and inspected, one by one. Because the slices are very thin, researchers focus on examining particular target regions. For example, one anatomist may make injections in a few sites in the parietal cortex and examine parts of the lateral prefrontal cortex for staining that indicates the presence of an anatomical connection. Injection by injection, study by study, neuroanatomists have compiled enough information to provide a good idea of the pathways crisscrossing the brain.

Although anatomical knowledge of pathways (and their strengths) is incomplete, the overall picture is one of *massive* connectivity. This is made clearer when computational analyses are used to combine the findings across a large number of individual studies. A field of mathematics that comes in handy here is called *graph theory*, which has become popular in the last two decades under the more appealing name "network science." Graphs are very general abstract structures that can be used to formalize the interconnectivity of social, technological, or biological systems. They are defined by *nodes* and the links between them, called *edges* (figure 10.1). A node represents a

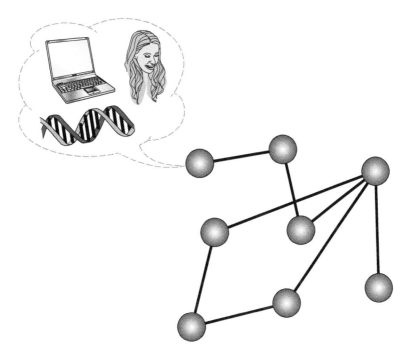

Figure 10.1
A graph is a mathematical object that can represent arbitrary collections of elements (persons, computers, genes), called nodes (denoted by circles), and their relationships, called edges (denoted by lines joining node pairs).

particular object: a person in a social group, a computer in a technological network, or a gene in a biological system. Edges indicate a relationship between the nodes: people who know each other, computers that are physically connected, or genes with related functions. So, in the case of the brain, areas can be represented by nodes, and edges interlinking them represent a pathway between them. (A so-called directed graph can be used if the direction of the pathways are known; for example, from A to B but not vice versa.)

Graph analysis demonstrates that brain regions are richly interconnected, a property of both cortical and subcortical regions. In the cortex, this property is not confined to the prefrontal cortex (which is often highlighted in this regard) but is observed for *all* lobes. Indeed, the overall picture is one of enormous connectivity, leading to *combinatorial* pathways between sectors. In other words, one can go from point A to point B in multiple ways, much like navigating a dense set of roads. Computational neuroanatomy has greatly refined our understanding of connectivity.

High global accessibility Rumors spread more or less effectively depending on the pattern of communication. They spread faster and farther among a community of college students than among faculty professors, assuming that the student community is more highly interconnected than professors are. This intuition is formalized by a graph measure called *efficiency*, which captures the effectiveness of information spread across members of a network, even those who are least connected (in the social setting, the ones who know or communicate the least with other members). How about the brain? Recent studies suggest that its efficiency is very high. Signals have the potential to travel efficaciously across the entire organ, even between parts not near each other and even between parts that are *not* directly connected; in this case, the connection is indirect, such as traveling through C, and possibly D, to get from A to B. The logic of the connectivity structure seems to point to a surprising property: physical distance matters *little*.

For many neuroscientists, this conclusion is surprising, if not counterintuitive. Their training favors a "processing is local" type of reasoning. After all, areas implement particular functions. That is to say, they are the proper *computational units*—or so the thinking goes (see chapter 4). This interpretation is reinforced by the knowledge that anatomical pathways are dominated by short-distance connections. In fact, 70 percent of all the projections to a given locus on the cortical sheet arise from within 1.5 to 2.5 millimeters (to give you an idea, parts of the occipital cortex toward the back of the head are a good 15 centimeters away from the prefrontal cortex). Doesn't this dictate that processing is local, or quasi-local? This is where math, and the understanding of graphs, helps sharpen our thinking.

In a 1998 paper entitled "Collective Dynamics of 'Small-World' Networks" (cited tens of thousands of times in the scientific literature), Duncan Watts and Steven Strogatz showed that systems made of locally clustered nodes (those that are connected to nearby nodes) but that also have a *small* number of random connections (which link arbitrary pairs of nodes) allow *all* nodes to be accessible within a small number of connectivity steps (Watts and Strogatz 1998).[2] We discussed this work in chapter 1, where we mentioned the idea of six degrees of separation. Starting at any arbitrary node, one can reach another (no matter which one) by traversing a few edges. Helping make the paper a veritable sensation, Watt and Strogatz called this property "small world." The strength of their approach was to show that this is a hallmark of graphs with such a connectivity pattern, irrespective of the type of data at hand (social, technological, or biological). Watts and Strogatz emphasized that the arrangement in question—what's called network topology—allows for enhanced signal-propagation speed, computational power, and synchronizability between parts. The paper was a game changer in how one thinks of interconnected systems.

In the 2000s, different research groups proposed that the cerebral cortex is organized as a small world. If correct, this view means that signal transduction between parts of the cortex can be obtained through a modest number of paths connecting them. It turns out that the brain is *more* interconnected than would be necessary for it to be a small world.[3] That is to say, there are *more* pathways interconnecting regions than the minimum needed to attain efficient communicability. So, while it is true that local connectivity predominates within the cortex, there are enough medium- and long-range connections—in fact, more than the "minimum" required—for information to spread around remarkably well.

Connectivity core ("rich club") A central reason the brain is *not* a small world is because it contains a subgroup of regions that is very highly interconnected. The details still are being worked out, not least because knowledge of anatomical connectivity is incomplete, especially in humans.

In 2010, the computer scientists Dharmendra Modha and Raghavendra Singh gathered data from over 200 anatomical tracing studies of the macaque brain (Modha and Singh 2010). Unlike most investigations, which have focused on the cortex, they included data on subcortical pathways, too (figure 10.2). Their computational analyses uncovered a "tightly integrated core circuit" with several properties: (*i*) It is a set of regions that is far more tightly integrated (that is, more densely connected) than the overall brain; (*ii*) information likely spreads more swiftly within the core than through the overall brain; and (*iii*) brain communication relies heavily on signals being communicated via the core. The proposed core circuit was distributed throughout the brain; it wasn't just in the prefrontal cortex, a sector often underscored for its integrative capabilities, or some other anatomically well-defined territory. Instead, the regions were found in all cortical lobes, as well as subcortical areas such as the thalamus, striatum, and amygdala.

In another study, a group of neuroanatomists and physicists collaborated to describe formal properties of the monkey cortex (Markov et al. 2013). They discovered a set of 17 heavily interconnected brain regions

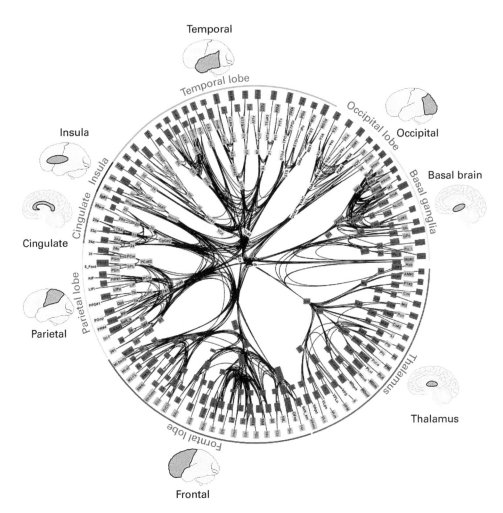

Figure 10.2
Massive interconnectivity exists between all brain sectors (shown in the brain insets). This computational analysis of anatomical connectivity was created by collating pathways (lines) from hundreds of studies. To improve clarity, only a subset of the connections is shown.
Source: Reproduced with permission from Modha and Singh (2010).

across the parietal, temporal, and frontal cortex. For these areas, 92 percent of the connections that could potentially exist between region pairs have indeed been documented in published studies. So, in this core group of areas, nearly every one of them can talk directly to all others—a remarkable property. In a graph, when a subset of nodes is considerably more well connected than others, it is sometimes referred to as a "rich club," in allusion to the idea that in many societies a group of wealthy individuals tends to be disproportionately influential.

Computational analysis of anatomical pathways has been instrumental in unraveling properties of the brain's large-scale architecture. We now have a vastly more complete and broader view of how different parts are linked with each other. At the same time, we must acknowledge that the current picture is rather incomplete. For one, computational studies frequently focus on cortical pathways. As such, they are cortico-centric, reflecting a bias of many neuroscientists who tend to neglect the subcortex (not to mention midbrain and hindbrain!) when investigating connectional properties of the brain. In sum, the theoretical insights of network scientists about "small worlds" demonstrated that signals can influence distal elements of a system even when physical connections are fairly sparse. But cerebral pathways vastly exceed what it takes to be a small world. Instead, what we find is a "tiny world."

Principle 2: High Distributed Functional Connectivity

A physical connection between two regions allows them to exchange signals, that much is clear. But there's another kind of relationship that we need to entertain—what we call a *functional connection*. Let's first consider an example unrelated to the brain, where in fact there aren't any physical connections. Genes are segments of DNA that specify how individual proteins are put together, and a protein itself is made of a long chain of amino acids. Proteins have diverse functions, including carrying out chemical reactions, transporting substances, and serving as messengers between cells. We can think of genes that guide the building of proteins that have *related* functions (for example, acting as hormones in the body, such as insulin, estrogen, and testosterone) as "functionally connected." The genes themselves aren't physically connected, but they are functionally related. In this section, we'll see how functional connectivity is a useful concept in the case of the brain.

At first glance, the notion of an architecture anchored on physical connections goes without saying. Region A influences region B because there is a pathway from A to B. However, the distinction between anatomy and function becomes blurred very quickly. Connections are sometimes "modulatory," in which case region A can influence the probability of responding at B, and sometimes connections are "driving," in which case they actually cause cells in B to fire. In many instances, the link between A and B is not direct but involves so-called interneurons: Region A projects first to an interneuron (often in area B itself), which then influences responses in other cells in B. The projections from interneurons to other cells in B can be excitatory or inhibitory, although they are often inhibitory. Of course, the strength of the fiber itself is critical. Furthermore, the presence of multiple feedforward and feedback pathways, as well as diffuse projections, further muddy the picture. Taken together, we see that connections between regions are not simply binary (they exist or not, as in a computer), and even a single weight value (say, a strength of 0.4 on a scale from 0 to 1) doesn't capture the richness of the underlying information.

Functional connectivity thus answers the following question: How *coordinated* is the activity of two brain regions that may or may not be directly joined anatomically? (See figure 10.3.) The basic idea is to gauge if different regions form a *functional unit*. What do we mean by "coordinated"? There are multiple ways to capture this concept, but the simplest is to ascertain how correlated the signals from regions A and B are. The stronger their

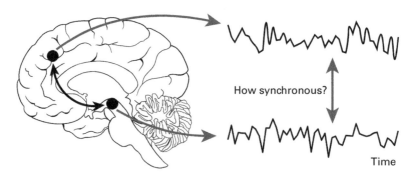

Figure 10.3
Functional connectivity measures the extent to which signals from two regions are in synchrony. Whether or not the regions are directly connected by an anatomical pathway is unimportant.

correlation, the higher the functional association or functional connection. Correlation is an operation that is summarized by values from −1 to +1. When two signals are perfectly related (which is never the case with noisy biological measurements), their correlation is +1; when they are in perfect opposition to one another (one is high when the other is low, and vice versa), their correlation is −1; when they are unrelated to each other, their correlation is 0 (this means that information about one of the signals tells us nothing about the other one, and vice versa).

Let's consider what I called the two-step property of the amygdala. Because this area is connected physically to approximately 40 percent of prefrontal subregions, it can influence a sizable portion of this lobe in a direct manner—that is, through a single step (such as regions in the orbitofrontal cortex and the medial prefrontal cortex). But approximately 90 percent of prefrontal cortex can receive amygdala signals after a single additional connection *within* prefrontal cortex (see Averbeck and Seo 2008). Thus, there are two-step pathways that join the amygdala with nearly all of the prefrontal cortex. Consequently, the amygdala can engage in meaningful functional interactions with areas that are not supported by strong direct anatomical connections (such as the lateral prefrontal cortex) or even not connected at all.

The foregoing discussion is worth highlighting because it is *not* how neuroscientists think typically. They tend to reason in a much more direct fashion, considering the influences of region A to be most applicable to the workings of regions B, to which it is directly connected—a type of connection called monosynaptic. To be sure, a circuit involving A → X → B is more indirect than A → B, and if the intermediate pathway involving X is very weak, the impact of A on X may be negligible. But the point here is that this needn't be the case, and we should not discard this form of communication simply because it is indirect (recall the discussion about network efficiency above).

It's natural to anticipate a functional association between brain regions that are directly connected. Yet the relationship between structural and functional connectivity is not always a simple one, which shouldn't be surprising because the mapping between structure and function in an object as interwoven as the brain is staggeringly complex. A vivid example of structure-function *dissociation* is illustrated by adults born without the corpus callosum, which contains massive bundles of axonal extensions joining the two hemispheres. Although starkly different structurally relative to controls, individuals without the callosum exhibit very similar patterns of *functional* connectivity compared to normal individuals (Tyszka et al. 2011). Thus, largely normal coordinated activity emerges in brains with dramatically altered structural connectivity, providing a clear example of how functional organization is driven by factors that extend beyond direct pathways.

The upshot is that to understand how behavior is instantiated in the brain, in addition to working out anatomy, it is necessary to elucidate the functional relationships between areas. Importantly, *anatomical* architectural features support the efficient communication of information, even when strong direct fibers aren't present, and undergird *functional* interactions that vary based on a host of factors.

An experiment further illustrating the above issues studied monkeys with functional magnetic resonance imaging (MRI) during a "rest" condition, when the animal was not performing an explicit task (Grayson et al. 2016). The researchers observed robust signal correlation (the signals went up and down together) between the amygdala and several regions that aren't connected to it (as far as we know). They asked, too, whether functional connectivity is more related to direct (monosynaptic) pathways or connectivity with multiple steps (polysynaptic) by undertaking graph analysis. Are there efficient routes of travel between regions even when they aren't directly connected? To address this question quantitatively, they estimated a graph measure called *communicability* (related to the concept of efficiency discussed previously), and they found that amygdala functional connectivity was more closely related to their measure of communicability than what would be expected by only considering monosynaptic pathways. In other words, polysynaptic, multistep routes should be acknowledged. In fact, their finding shows that to understand the relationship between signals in the amygdala and that of any other brain region, it's important to consider all pathways that can bridge them.

Principle 3: Networks as Functional Units

In a highly networked system like the brain, we need to shift from thinking in terms of isolated brain regions and adopt the language of networks: Networks of brain regions collectively support behaviors. *The network itself is the unit*, not the brain area (figure 10.4). Consequently, processes that

Region   Network

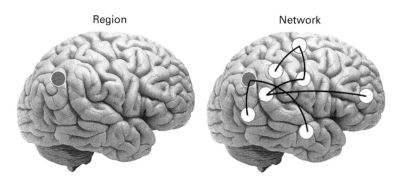

Figure 10.4
What's the rightful functional unit of interest? Historically, the brain area took center spot. A better unit is a network of brain regions working together.

support behavior are not implemented by an individual area but depend on the interaction of multiple areas, which are dynamically recruited into multiregion assemblies (more on dynamic aspects below).

These ideas are now new. In 1949, Donald Hebb proposed that the brain's ability to generate coherent thoughts derives from the spatiotemporal orchestration of neuronal activity. He hypothesized that a discrete, strongly interconnected group of active neurons, the *cell assembly*, represents a distinct mental entity. Hebb conceptualized the cell assembly as quite distributed across the brain, at least in some cases; for example, in one case, the neuronal coalition involved cells in the cortex and thalamus and possibly the basal ganglia (Hebb 1949, xix). But the exact spatial extent of a cell assembly was probably not as important because he believed they could be organized into "systems of assemblies"—in other words, larger ensembles that would involve more neural real estate. The concept of a cell assembly as a spatially distributed unit spurred many theoretically inclined neuroscientists to search for them, providing the conceptual seed of many mathematical models of neural computation.

How should we think of brain networks? Cell recordings are typically constrained to particular brain regions, so they don't provide information about distributed circuits. Hence, not surprisingly, information about networks has originated from other techniques, such as functional MRI. Although this recording modality provides only indirect measures of neuronal activation, it has uncovered the existence of large-scale networks. Their existence can be illustrated with what is called "seed" analysis, where activity from one region is correlated with signals at other locations (remember that in functional MRI, activity from the entire brain is usually collected). The general goal is to determine the extent to which each brain location is synchronized with the seed—in other words, the strength of their functional connectivity. If we now consider the regions that are most strongly correlated with the seed region, we can call this set (including the seed) a *functional network*.

This approach summarizes some of the main ingredients of identifying large-scale networks in the brain using functional MRI. Employing more formal and systematic methods ("seeding" each brain region individually would be very inefficient), researchers have described multiple networks. For example, based on data from a thousand participants, Thomas Yeo and colleagues (Yeo et al. 2011) subdivided the entire cortex into seven networks, including what they called the "frontoparietal" network with regions in the frontal and parietal cortex, and the "visual" network spanning the occipital and temporal cortex (figure 10.5). An intriguing network is called "default," based not on the original meaning of the word ("failure" or "failure to act") but on the computer science sense of "selecting automatically in the absence of a choice made by the user." The "user" in this case is the participant being scanned. But what about the "absence of choice"?

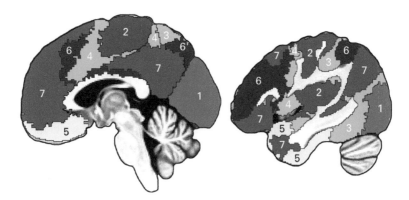

Figure 10.5
Large-scale networks revealed by functional MRI data. Each gray level (as well as color level) represents a specific network (also called "community"), such as the "frontoparietal" (indicated by the number 6). Signals measured across all locations in a network are relatively synchronized, suggesting that they work in a coordinated fashion.
Source: Networks from Yeo et al. (2011).

Naturally, the regular way an experiment takes place is by instructing participants on what to do—for example, try to remember a list of words, pay attention to certain types of stimuli, imagine something, move a finger. But then there was the suggestion of scanning the person when asking her to do nothing, as in "think of nothing in particular." What an idea! Surprisingly, a set of brain regions during this "no explicit task" condition exhibits strongly synchronized activity, providing the basis to define the "no-task network." However, the name "default" was suggested instead and caught on.

What might be the function of this network? Somehow, the question has proved irresistible and a large literature has sprawled to try to answer it. When someone is not explicitly required to perform a task, more likely than not, they will entertain themselves with self-relevant events: thinking about where they need to go after the scan, remembering an unpleasant event over the past weekend, anticipating a set of exams during finals week, and so on. Thus, the retrieval of biographically related memories, and the anticipation of future events, probably takes center stage. Of course, someone could use the time to try to memorize irregular subjunctive verbs they are learning in Italian, or go over the proof of a mathematical theorem they were just shown in a class, but that's perchance the exception. By and large, one's mentation will be dominated by "self-related processing." So, one of the main ideas is that the default network is heavily preoccupied with "self-processing."

Evidence of large-scale networks is found with other techniques, too. This is the case with surface electroencephalogram (EEG) recordings in humans (when electrodes are placed on the surface of the cortex in patients being prepared for neurosurgery), as well as with electrodes inserted across different parts of the brain simultaneously in animals. But given the limited spatial coverage of these techniques, in most cases the emphasis is on pairwise interactions, such as cortical-cortical interactions between parts of the parietal and frontal cortex, and subcortical-cortical interactions, such as between the prefrontal cortex and the thalamus. With the development of new techniques, it should be possible in the near future to address broad-range functional properties of brain signals with techniques that measure neuronal activity signals more directly.

Principle 4: Interactions via Cortical-Subcortical Loops

Computational analysis of anatomical datasets tells us a story of massive combinatorial connectivity enriched by properties like high global accessibility. But thus far this work has missed major features of the connectional architecture of the brain uncovered by traditional anatomical research, the one involving the laborious tracing of pathways. This is particularly the case because much of the computational work has centered on cortical data. If we think of connectivity along the cortex as "horizontal," a cortico-centric standpoint misses the "vertical" (cortical-subcortical) features of anatomical pathways. Here, I will build on chapter 9's discussion in of loops that unite the cortex with the subcortex.

The striatum is a rather oddly shaped structure at the base of the forebrain. The entire cortical sheet (with the exception of primary visual cortex) projects to this area. In case the reader failed to appreciate the scope of this statement, I'll repeat it: But for one small piece, *all* of the cortex projects to the striatum. Although neuronal connections are often bidirectional, the striatum does not directly reciprocate the pathways it receives. But the region is not a connectivity sink. In the 1980s, it was found that the striatum loops back to the cortex in an intriguing way through the thalamus: The cortex projects to the striatum, which projects to the thalamus, which by its turn projects back to (roughly) the same regions that send fibers to the striatum—a loop is formed (see figure 9.6).

Cortical-subcortical loops are just one example of a major connectivity system that isn't adequately captured by computational analyses of anatomical data. Whereas other examples could be discussed, the take-home message is that the extensive cortical-subcortical connectivity substantially extends the brain's communication architecture *beyond* that considered in principle 1. Indeed, signal interchange and integration are likely vastly more complex than currently fathomed.

Let's discuss the "cortico-basal ganglia-thalamo-cortical" loops (quite a mouthful), which I'll refer to it as the cortical-basal ganglia loops. Are the circuits "open" or "closed"? In other words, do cortical regions projecting to the striatum receive feedback from the same striatal districts (via the thalamus)? This type of organization constitutes a closed loop, with an open loop being closer to an arrangement in which return projections target other cortical areas.

Theoretically, the possibility of closed loops requires a fine level of precision, especially when a multistep pathway is at stake. Each connection must be arranged *topographically*: Neighboring points in area A project to correspondingly neighboring points in area B. Topography, therefore, speaks to *specificity*. Consider other types of arrangement: A local patch in area A projects to a spatially diffuse patch in area B, or even to multiple areas B, C, and D. Topographically organized connectivity allows a certain degree of territorial segregation, thus creating processing streams that maintain relative independence from others.

There is some evidence for a closed-loop organization of the basal ganglia circuit. Different parts of the cortex project to different sectors of the striatum, which project back to the originating cortical areas. In fact, the observation of topography led to the classification of loops based on the part of the cortex projecting to the striatum (and receiving return projections), particularly in the frontal cortex: "motor," "oculomotor," "dorsolateral prefrontal," "lateral orbitofrontal," and "anterior cingulate." Although this way of subdividing basal ganglia loops became popular, there is ample crosstalk between them, too (Shipp 2017). While some circuits are more closed-loop (such as the one involving motor cortex), others form more of an open-loop circuit (such as those projecting to the ventral parts of the striatum). We don't find a single type of organization but a spectrum of arrangements.

Basal ganglia loops are by far the most often emphasized cortico-subcortical connectional system. Perhaps because of their arrangement in terms of loops, they gained notoriety in ways that other systems did not. But additional large-scale connectivity systems involve other structures at the base of the forebrain, too, perhaps most notably the amygdala.[4] The cerebellum (in the hindbrain), which is still rather poorly understood, is also densely interconnected with the cortex, as well as with the basal ganglia itself. Again, giant webs of connectivity can form.

Principle 5: Connectivity with the Body

We think of the brain as controlling our movements and accompanying sensations through the musculoskeletal system. The sensory part includes both tactile impressions and proprioception, which is our sense of the position of our body in space (when doing a headstand, we feel that we are upside down). But it's easy to forget that the central nervous system is in constant two-way communication with the body on a much broader scale through a process called *interoception* (see chapter 6) that is both varied and nuanced, encompassing feelings related to muscular and visceral sensations, pain, and itch, among many others.

Interoception heavily depends on the brainstem and upper cervical cord. These sectors receive substantial bodily messages and pass them along expediently to the thalamus, where they reach the insula in the cortex. They also reach the hypothalamus (in the basal forebrain), which is a key node in the regulation of the internal state of the body and, as such, is part of circuits that maintain life (chapter 5). The hypothalamus is bidirectionally connected with almost all of the cortex, too. In principle, the state of the body can influence most, if not all of, the entire brain, and vice versa.

How does the brain influence the body? In the 1940 and 1950s, electrical stimulation of the cortex of patients being prepped for neurosurgery became routine, leading to the discovery of regions with clear impact on the body state (chapter 6). Changes in respiration, blood pressure, heart rate, and the diameter of the pupil were routinely detected. Electrical stimulation of the cingulate cortex, for example, impacts virtually all autonomic processes as well as many endocrine mechanisms. Many of these effects are carried out through the hypothalamus, although direct projections from the cortex to the medulla lead to even more immediate influences on the body.

The brain keeps the body alive, but it can cause injury, too. Chronically enhanced heart rate is a risk factor for premature death, as is reduced heart rate variability (which is simply a measure of the fluctuation in time between each heartbeat). Julian Thayer and colleagues have documented how both experimentally induced and dispositional measures of worry are associated with high heart rate and low heart rate variability (Brosschot, Gerin, and Thayer 2006; Brosschot, Verkuil, and Thayer 2018). Put simply, worry kills you (see chapter 5).

What Kind of Functional Networks?

Functional networks are based on the relationships of the signals in disparate parts of the brain, not on the status of their physical connections. The spatial scale of functional circuits varies considerably, from those linking nearby areas to large ones crisscrossing the brain. The most intriguing networks are possibly those discovered with functional MRI. To identify networks, investigators capitalize on "clustering methods," general computer science algorithms that sort basic elements (here, areas) into different groups. The objective is to subdivide a set of elements into natural clusters, also known as *communities*. (These are also called modules by network researchers, but this is confusing in the case of neuroscience given the meaning of "modularity" discussed in chapter 4.) Intuitively, a community should have more internal than external associations. For example, if we consider the set of all actors in the United States, we can group them into theater and film clusters (theater actors work with and know each other more so than they work with and know film actors). This notion can be formalized: Communities are determined by subdividing a set of objects by maximizing the number of *within*-group connections and minimizing the number of *between*-group connections. Remember that a connection in a graph is a link between two elements that share the relationship in question, such as between two theater actors who have worked together or two actors who were in the same movie. Thus, theater actors will tend to group with other theater actors and less so with film actors, and vice versa.

The most popular partitioning schemes parse individual elements (brain regions in a brain network, persons in a social network, etc.) into unique groupings—a node belongs to one and exactly one community. Based on functional MRI data at rest, the study by Yeo and colleagues discussed above described a seven-community division of the entire cortex, where each local patch of tissue belongs to a single community. In other words, the overall space is broken into *disjoint* communities. Their elegant work has been very influential, and their seven-network partition was adopted as a sort of canonical subdivision of the cortex (see figure 10.5). (Intriguingly, they also described an alternative 17-community subdivision of the cortex, but this one didn't become very popular, likely because 17 is "too large" for neuroscientists to wrap their heads around.) Whereas discrete clusters simplify the description of a system, are they too inflexible, leading to the loss of valuable information?

Think again about the community of actors. Perhaps they neatly subdivide into theater and film groups and perhaps into some other clear subgroups, such as Broadway and Off-Broadway theater performers. Yet real-world groupings are seldom this simple, and in this case a particular artist might belong to more than one set (acting in both theater and film, say). In fact, several scientific disciplines, including sociology and biology, have realized the potential of *overlapping* network organization. For example, the study of chemical interactions reveals that a substantial fraction of proteins interact with several protein groups, indicating that actual networks are made of interwoven sets of overlapping communities (Palla et al. 2005, 814). How about the brain?

Consider the versions of the connectivity "core" discussed previously, which contains regions distributed across the cortex or across the entire brain. These areas are not only strongly *inter*connected but also linked with many other parts of the brain. Put another way, by definition, regions of the core are those talking to lots of other ones. Traditional disjoint network partitioning schemes emphasize the *within* community grouping while minimizing the *between* community interactions. But regions of the core are both very highly interconnected and linked with disparate parts of the brain. So, how should we think about them? Network science has additional tools that can help. One of them is to think of nodes as having a spectrum of computational properties. Both how well connected a node is and how vastly distributed its links are matter. Nodes that are particularly well connected are called *hubs* (with a meaning similar to that in "airport hub"), a property that is formally captured by a mathematical measure called *centrality*. Hubs come in many different flavors, such as *connector hubs* that have links to many communities and *provincial hubs* that are well connected within their particular community (Guimera and Nunes Amaral 2005). We can thus think of connector hubs as nodes that are more "central" in the overall system than provincial nodes.

Nodes that work as connector hubs are distinctly interesting because they have the potential to integrate diverse types of signals (if they receive connections from disparate sources) or to distribute signals widely (if they project to disparate targets). They are a particularly good reminder that communities are not islands; nodes within a community have connections both within their community and to other clusters.

We suggested that a better brain unit is a network, not a region. But in highly interconnected systems like the brain, subdividing the whole system into discrete and separate networks still seems too constraining. (The approach is more satisfactory in engineering systems, which are often designed with the goal of being relatively modular.) An alternative is to consider networks as inherently *overlapping*. In this type of description, collections of brain regions—networks—are still the rightful unit, but a given region can participate in several of them, like in the example of the actors community discussed previously. Thinking more generally, we can even describe a system in terms of communities but allow every one of its nodes to belong to *all* communities, simultaneously. How would this work? See figure 10.6.

In a study in my lab, we allowed each brain region to participate in a community in a graded fashion, which was captured by *membership values* varying continuously between 0 and 1, with 0 indicating that the node did not belong to a community and 1 indicating that it belonged uniquely to a community (Najafi et al. 2016). One can think of membership values as the strength of participation of a node in each community. It's also useful to conceive of membership as a finite resource, such that it sums to 1. For example, in the case of acting professionals, a performer could belong to the theater cluster with membership 0.7 and to the film cluster with membership 0.3 to indicate the relative degree of participation in the two. In my lab's study, we found that it was reasonable to subdivide cortical and subcortical regions into five, six, or seven communities, like other algorithms have suggested in the past. But we also uncovered dense community overlap that was not limited to "special" hubs. In many cases, the entire community was clearly a meaningful functional unit, while at the same time most of its nodes still interacted nontrivially with a large set of brain regions in other networks.

The results of our study, and related ones by other groups, suggest that densely overlapping communities are well suited to capture the flexible

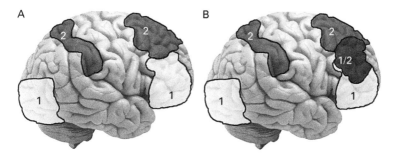

Figure 10.6
Disjoint and overlapping networks. (a) Disjoint partitioning is most commonly studied in neuroscience. (b) Brain networks can also be conceptualized as overlapping. In this case, areas can belong to multiple networks, simultaneously, such as the patch labeled "1/2" that belongs to both communities.

and task-dependent mapping between brain regions and their functions. The upshot is that it is very difficult to subdivide a highly interconnected system without losing a lot of important information. What we need is a set of tools that allow us to do this in sophisticated ways. And we need them both to think about how networks are organized in space, as discussed in this section, and in time, to which we turn next.

Networks Are Dynamic

The brain is not frozen in place but is a dynamic, constantly moving object. Accordingly, its networks are not static but evolve temporally. As an individual matures from infancy to adolescence to adulthood and old age, the brain changes structurally. But the changes that I want to emphasize here are those occurring at much faster timescales, those that accompany the production of behaviors as they unfold across seconds to minutes.

Functional connections between regions—the degree to which their signals covary—are constantly fluctuating based on cognitive, emotional, and motivational demands.[5] When someone pays attention to a stimulus that is emotionally significant (it was paired with mild shock in the past, say), increased functionally connectivity is detected between the visual cortex and the amygdala. When someone performs a challenging task in which an advance cue indicates that they may earn extra cash for performing it correctly, increased functional connectivity is observed between the parietal/frontal cortex (important for performing the task) and the ventral striatum (which participates in reward-related processes). And so on. Consequently, network functional organization must be understood dynamically. In the past decade, researchers have created methods to delineate how networks change across time, informing how we view social, technological, and biological systems.

Brain networks are dynamic. For example, the frontoparietal network mentioned previously is engaged by many challenging tasks, such as paying attention to an object, maintaining information in mind, or withholding a prepotent response. If a person transitions mentally from, say, listening passively to music to engaging in one of these functions (say, needing to remember the name of a book just recommended to them), the state of the frontal-parietal network will correspondingly evolve, such that the signals in areas across the network will increasingly synchronize, supporting the task at hand (holding information in mind).

There's a second, more radical way in which networks are dynamic. That's when they are viewed not as fixed collections of regions but instead as coalitions that form and dissolve to meet computational needs. For instance, at time t_1, regions R_1, R_2, R_7, and R_9 might form a natural cluster; at a later time t_2, regions R_2, R_7, and R_{17} might coalesce. This shift in perspective challenges the notion of a network as a coherent unit, at least for longer periods of time. At what point does a coalition of regions become something other than community X? For example, the brain areas comprising the frontoparietal network exist irrespective of the current mental operation; for one, the person could actually be sleeping or under anesthesia. The areas in question may not be strongly communicating with each other at all. Should it be viewed as a *functional unit*? When the regions become engaged by a mental operation, their signals become more strongly synchronized. But when along this process should the network be viewed as "active"? As the mind fluctuates from state to state, we can view networks cohering and dissolving correspondingly—not unlike a group of dancers merging and separating as an act progresses. The temporal evolution of their joint states is what is important.

Large-Scale Functionally Integrated Systems

Researchers who study humans frequently focus on circuits centered on the cortex—for example, involving regions of the parietal and frontal cortex that are important for attention. On the other hand, investigators who focus on nonhuman animals like mice and rats tend to focus on non-cortical circuits—for example, including regions of the striatum and midbrain that are important for motivated behaviors, such as performing a task that leads to a rewarding food morsel. But the brain doesn't obey the boundaries imposed by investigators. Multiple levels along the neuroaxis work together.

Let's consider the following classes of circuits (figure 10.7):

1. Cortical circuits
2. Subcortical/brainstem circuits
3. Cortical-subcortical loops
4. Descending systems
5. Ascending systems

Cortical circuits The cortex is very richly interconnected. Not only do we encounter pathways between nearby regions (such as primary and

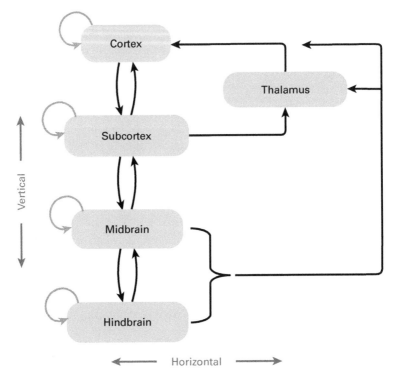

Figure 10.7
Horizontal and vertical communication in the central nervous system. The cortex
and basal subcortex are part of the forebrain, and so is the thalamus. The midbrain
and hindbrain are situated in the brainstem.

secondary visual cortex) but between distant regions, too (say, uniting temporal and frontal cortex). As discussed in chapter 2, cortical areas are physically linked by organized white matter fiber tracts.

Subcortical/brainstem circuits This group is considerably heterogeneous
and contains three sectors: forebrain, midbrain, and hindbrain. Multi-area
circuits exist within all these levels but interconnect them, too. For example,
amygdala projections traverse downward along the brainstem establishing
synapses at multiple sites, several of which project back, thus forming circuits that engage multiple areas. In particular, the amygdala is bidirectionally connected with both the ventral tegmental area and the periaqueductal
gray in the midbrain, which are themselves connected.[6]

Cortical-subcortical loops Cortico-basal ganglia circuits are the paradigmatic example here; they involve nearly all of the cortex. Another substantial connectional system interlinks the amygdala with cortex; in this case, some pathways return to the cortex through the thalamus, like in the case of basal ganglia loops, but there are abundant direct projections from the amygdala to the cortex, too.

Descending systems Some subcortical structures at the base of the forebrain project along the entire extent of the brainstem. This is illustrated, as mentioned, by the amygdala with its connections from the midbrain tegmentum to the spinal cord. Likewise, the hypothalamus projects throughout the brainstem.

Ascending systems Multiple brainstem areas synthesize neurotransmitters that are propagated across the brain, including the subcortex of the forebrain and the cortex.

How do all of these systems interact? Here, it is useful to think in terms of "horizontal" and "vertical" interactions and circuits (see figure 10.7). In the horizontal dimension, the main elements are at the same level of the brain, such as those involving cortex-to-cortex pathways. The vertical dimension crosses levels, such as in basal ganglia loops. Overall, the horizontal and vertical dimensions support the communication and integration of signals across varying spatial extents. In addition, relatively closed and relatively open circuits provide complementary designs. The former provide more segregation of the processing stream from outside influences; the latter support the distribution of signals giving rise to wide-ranging effects and the blurring of boundaries and categories—perception versus action, cognition versus emotion, and so on. Circuits join the cortex with the subcortical forebrain, the subcortical forebrain with the brainstem, and all of them together (figure 10.8), leading to multiple convergence regions—hubs—across the neuroaxis.

The domain of anatomy provides the necessary structural backbone necessary for communication; yet, out of it, a functional circuit is momentarily lifted, sculpted out of the anatomy. The particular circuit comprises populations of neurons that are coactive but spatially distributed. In this sense, it is not spatially contiguous but functionally coherent. Thus, although

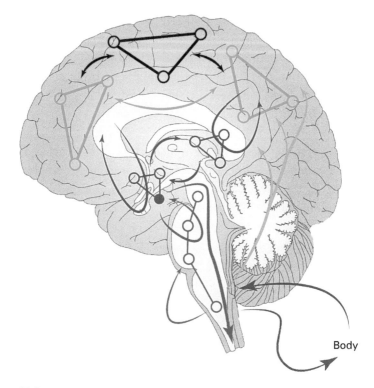

Figure 10.8
Functionally integrated systems link multiple levels of the brain, such that local circuits combine with larger and larger circuits.

function is always anchored on anatomy, circuits should be understood at the level of functional relationships.

I propose that large-scale connectional systems, according to the ideas developed in this chapter, are critical for understanding how complex behaviors are instantiated by the brain. I call them *functionally integrated systems* (figure 10.8).[7]

The (Misguided) Search for the Emotional Brain

As an illustration of how the concept of functionally integrated systems is useful in relating structure and function, let's consider the search for the neuronal underpinnings of emotion—the search for where emotion resides in the brain. Neuroscience has chased this question for a century and a half.

Emotions mobilize the body through autonomic, neuroendocrine, and musculoskeletal systems, in part through functionally integrated systems that have access to the hypothalamus and to structures in the brainstem and medulla that are linked to the body. Emotions also mobilize disparate brain responses, influencing attention, memory, and decision making. The engagement of body and brain, which is closely associated with neurotransmitter systems, relies initially on the most robust anatomical pathways but is rapidly expanded to include vast portions of the brain. The way this is accomplished follows from the general organizational principles described in the beginning of the chapter, including massive combinatorial anatomical connectivity and distributed functional connectivity. These properties, in conjunction with substantial network overlap, ensure that events of biological significance lead to the temporal evolution of network structure to meet the demands faced by the organism.

Emotion is at times likened to a "biasing" mechanism, such as directing perception to focus on a particularly relevant object, or shifting cognition from one type of information to another. Emotion is not adequately captured by this idea—it's much more. Emotion dynamically influences the properties of large-scale networks, including those that are described as perceptual, motor, motivational, or cognitive.

The proposal helps clarify, too, why some structures are so important for emotion, such as the amygdala and the hypothalamus—they are important hubs of distributed functionally integrated systems. Adoption of the framework bares naked the shortcomings of pointing to specific areas as constituting the "emotional brain" or even to specific levels of the brain, as in the focus on the cortex of some human research and the focus on the subcortex in some animal work. Ultimately, emotion—insofar as it is meaningful to speak of "emotion"—like every other mental domain, is a large-scale network property of the nervous system.

11 Unlearning Fear

In chapter 10, we discussed how brain circuits are key to generating complex behaviors in general terms. But how does it come about in practice? To provide a more concrete example, here we discuss *extinction learning*: After learning the association between a conditioned stimulus (say, a light) and an unconditioned stimulus (say, a shock), how does an animal learn that a light no longer predicts shock when the world has now changed and the one no longer leads to the other?

Much like other nine-month-olds, "little Albert" was not really bothered by the presence of a small white rat. Unlike other infants, however, little Albert was a subject in a study by John Watson, one of the early proponents of behaviorism. In one of the most controversial experiments in psychology, Watson and his assistant, Rosalie Rayner, applied their knowledge of classical conditioning to induce fear in the infant. They presented the boy with a white rat and then loudly clanged an iron rod. Not surprisingly, little Albert responded by crying. After multiple paired presentations, Watson and Rayner presented the white rat by itself, which led to a "fear response" (the boy cried). They had conditioned an initially neutral stimulus, which now evoked a response originally triggered by the loud noise. Priding himself on his ability to shape people's emotions, Watson later went into advertising and published an influential book on infant psychological care—yes, gasp.

In what now can only be seen as a perverse experiment, Watson and Rayner were building on knowledge about classical conditioning, most notably on the experiments of the Russian physiologist Ivan Pavlov. Today, in school, we all learn about Pavlov's dogs and how they came to salivate on hearing the bell. His discovery of the conditioned response was one of his most significant contributions to physiology and psychological science. (Pavlov earned the Nobel Prize in 1904 "in recognition of his work on the physiology of digestion, through which knowledge on vital aspects of the subject has been transformed and enlarged.")[1] Pavlov was also very interested in what he called the "internal inhibition of conditioned reflexes." He noted that the absence of reinforcement resulted in a weakening or disappearance of acquired behaviors; for example, in dogs, the discontinuation of food delivery on hearing the bell led to the weakening of salivation. More generally, when a conditioned stimulus (CS) no longer predicts the unconditioned stimulus (UCS; say, a shock) to which it was paired in the past, the CS gradually stops eliciting the conditioned response. This process is called *extinction*.

Fear conditioning has been among the most influential paradigms in all of psychology. Whereas extinction has not been so popular, it has also attracted a lot of attention. The importance of both phenomena, which include an extended family of related paradigms, transcends the laboratory, of course. Four out of five Americans will be exposed to a trauma during their lifetime, and many of them will develop a form of anxiety disorder, including phobia, social anxiety, and posttraumatic stress disorder. These conditions can be extremely debilitating and substantially impair quality of life. Not surprisingly, an array of psychological therapies has been developed to try to cure or at least ameliorate these conditions.

Consider trauma-focused approaches, such as "exposure therapy." When people are highly fearful of something, they tend to avoid the feared objects, activities, or situations. A person may avert parties if they experience social anxiety, for instance. Avoidance is initially beneficial; after all, the feared object or situation induces feelings that can be very unpleasant. But in the long term, avoidance can have rather negative consequences by excluding the person from the feared objects/situations. In exposure therapy, one is exposed to the fear-triggering event but in a safe environment. The idea is that if the event is not followed by an aversive experience, after multiple such pairings, the individual will be desensitized and the event will no longer evoke an unpleasant outcome. The general logic of exposure therapy is therefore that of the extinction processes. So, a CS no longer followed by a UCS *extinguishes* the conditioned response, which is the one that is so negative. When exposure therapy works, what makes it successful? When does it work best? At present, we don't know the answers to these questions because extinction has turned out to be fiendishly challenging to dissect, both at the behavioral and the neuroscientific levels.

Fear Extinction as Inhibition of Emotion by Cognition

Let's delve more deeply into the mechanisms of *fear extinction* (figure 11.1). When a conditioned stimulus no longer predicts the unconditioned stimulus to which it was paired at some point in the past, the CS stops eliciting the conditioned response. Pavlov himself believed this change involved the "development of internal inhibition," but was quite vague about the mechanisms involved, aside from alluding to cortical cells "entering into a state of inhibition" (Pavlov 1927). But the notion that the CS acquires *inhibitory* properties that allow it to suppress the conditioned response has played a central role in thinking about extinction.

The *acquisition* of fear itself during classical conditioning relies on the amygdala as well as several of its targets in the brainstem (chapter 5). But how about extinction? The role of the prefrontal cortex in the regulation of behavior in general, and emotion in particular, is a perennial theme in neuroscience. If investigating the question in a rat, a natural candidate to look at would be the medial sector of the frontal lobe, as rats don't have a prominent lateral prefrontal cortex (some researchers even question whether they have brain areas that are comparable to the lateral prefrontal cortex of primates). In the early 1990s, Joseph LeDoux and colleagues reported that

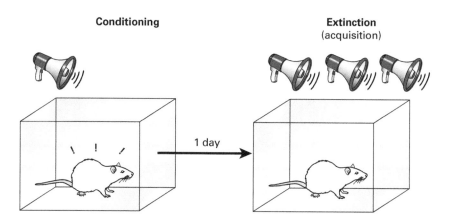

Figure 11.1
Unlearning fear. After aversive conditioning, the conditioned stimulus is presented alone multiple times until a conditioned response is no longer produced. Whereas after conditioning the animal freezes after the tone, once extinction occurs, the animal moves around normally.

the medial prefrontal cortex plays an important role during fear extinction (Morgan, Romanski, and LeDoux 1993). Animals with a lesion of the medial prefrontal cortex (PFC) took considerably longer to extinguish learned associations. Because the medial PFC is extensively interconnected with the amygdala, as well as with several of its brainstem targets that generate conditioned responses, the findings resonated with the idea that the medial PFC inhibits the amygdala, thereby halting the conditioned response. This mechanism of fear extinction fit the old formula: cognition, tied to the medial PFC, controlling emotion, itself tied to the amygdala and other subcortical structures (figure 11.2).

Behaviorally, what is extinction? In the experiment by LeDoux's group, following the extinction procedure, they tried to determine the general fearlessness of the rat. Had the animals simply become unusually bold and the presumed fear extinction a manifestation of their new personality? No, the general fearfulness of rats didn't seem altered. But when presented with the prior CS, they didn't mind it as much as before, and so the stimulus's ability to produce freezing behavior was diminished.

Let's delve deeper into the extinction process. When the CS no longer predicts an aversive outcome, it behooves the animal to take into account

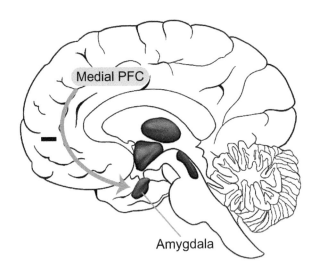

Figure 11.2
Early proposed mechanism for fear extinction. The medial prefrontal cortex (PFC) inhibits the amygdala, thereby preventing the conditioned response from being generated when the conditioned stimulus no longer predicts it.

this information, learning features of the now-safe environment that signal safety. But when the CS no longer predicts the UCS, why exactly is this case? Is the CS being presented in a completely different environment or context? What if the original CS (call it CS1) now appears simultaneously with another environmental stimulus (call it CS2). If no UCS ensues, should safety be deemed thanks to the CS1 (perhaps it no longer predicts the UCS) or to the CS2 (CS2's appearance now makes the world safe)? Experiments show that if, during the extinction process (when the CS1 is presented without the concomitant UCS), another stimulus (CS2) is presented alongside the original CS1 will *not* be treated as safe. This situation is at times called "protection from extinction." In other words, the relationship between the original CS and the UCS is *maintained*, and when the CS is presented alone, it produces a conditioned response—"fear" continues. The absence of the UCS is being attributed to the additional factor (the CS2), and the animal had better be careful (about CS1). A similar protection from extinction takes place when a new action concurrently performed by the animal leads to safety (that is, prevents the occurrence of the UCS). Here, the action is attributed with the power to ward off the punishment. So the animal will still fear CS1.

The intelligence of the learning processes is further highlighted by a scenario called "backward blocking." Suppose a compound stimulus, CS1 + CS2 (such as a light and a tone), is associated with a UCS. Once learning occurs, by definition the presentation of the pair CS1 + CS2 will generate a conditioned response. When either CS1 or CS2 is presented alone, some amount of conditioned responding ensues, although weaker than when the pair is presented. Now, if from this point forward one of the elements of the compound stimulus (say, the light) is consistently paired with the UCS, the second element (the tone) will *cease* to generate a conditioned response. It seems that since the light can fully predict the UCS, the tone is regarded as unrelated to the UCS, indeed *retrospectively*. This is all the more striking because the predictive value of a CS can be updated despite its absence. In the present case, the value of the tone is updated when the light is presented by itself. Whatever extinction is, it's not dumb!

In sum, extinction is more than a simple form of inhibition. It is a sophisticated form of *learning*, and as such the formation of an "extinction memory" involves processes akin to those observed in learning in general: acquisition, consolidation, and retrieval. What is being learned is safety.

The manner by which this memory influences behavior depends on how it was established (acquisition), how it was strengthened (consolidation), and how it will be reactivated (retrieval) in particular situations. The chief goal is to learn what should be feared, and therefore avoided, and what is safe and doesn't call for special measures and might be even approached. The factors that drive this process fall into two categories: those that promote defensive responding (such as "fearing" the stimulus) and those that do not ("safety" responding). To successfully accomplish extinction, the nature of the CS-UCS relationship needs to be unraveled: the CS might no longer predict the UCS, the CS might predict the UCS less reliably than before, or the CS might predict the UCS just as well as before, but something else is preventing the UCS from occurring. The conclusion that is favored will determine if the animal will express fear or not on encountering the stimulus in the future.[2]

Diving Deeper into the Mechanisms of Extinction

We saw that the medial prefrontal cortex plays an important role during extinction. Given that the region is extensively interconnected with the amygdala, it was natural to think that the former inhibits the latter. Nevertheless, since the early studies in the mid-1990s, the picture that the medial PFC *controls* the amygdala has been muddied considerably. For example, chemical blockage of the basolateral amygdala either impairs or entirely prevents extinction in the first place.[3] Furthermore, morphological changes in synapses in the amygdala itself support the consolidation of extinction. These findings strongly counter the notion that the amygdala is simply inhibited by the medial prefrontal cortex. Instead, it is a critical site for the *formation* of safety memories, very much like it is important for processes that establish fear memories themselves.

Anatomically, the amygdala isn't only the target of pathways from the medial PFC but also projects *to* it. Together with the findings above, it becomes untenable to place the amygdala as "down" from the medial PFC, and in fact some investigators propose that the amygdala actually should be viewed as "upstream" of the medial PFC.[4] Put another way, the amygdala and the medial PFC interact in complex ways during extinction. It is well established that the amygdala plays a critical role in establishing the association between a CS and a UCS (fear learning). It is increasingly clear that it participates in a major way in learning safety (extinction), too.

When a CS no longer predicts a UCS, the specific environment where extinction learning takes place is paramount. The animal learns that the CS *in this environment* is now safe. Indeed, if the CS now reappears in a novel context, the animal displays defensive behaviors—the CS does not signal safety there. Studies have shown that the hippocampus keeps track of the context in which extinction occurs. This type of learning is essential; after all, it could be disastrous for the animal to generalize the safety of a CS to situations unlike those from where extinction took place. The contributions of the hippocampus to contextual learning will be discussed below, but first let's consider some of the functions of this structure—some of which were learned the very hard way.

Hippocampus: A Brief Detour into a Tragic Neurosurgery

At the age of 27, Henry Molaison (known as patient HM when he was alive) was referred to William Scoville, a neurosurgeon at Hartford Hospital. Despite maximum medication of various forms, he could not lead a regular life.[5] He had worked for a time at an assembly line as a motor winder but had become so incapacitated by his seizures that working was no longer possible. The year was 1953, and a radical clinical approach was taken. With the understanding and approval of the patient and his family, the "frankly experimental" operation was undertaken. The surgery didn't eliminate the seizures, but they became less debilitating than before. A standard IQ test indicated that his score was unaltered compared to the test taken before surgery. His personality also appeared stable. So far, so good. Yet, Molaison exhibited a profound, indeed devastating, memory impairment. For example, just before his psychological examination in April 1955, he had been talking to a famous neuroscientist, Karl Pribram, but he formed no memory of this event and denied that anyone had spoken to him. During conversations, he constantly recounted boyhood events and, eerily, didn't appear to realize that he'd had an operation. Something was clearly amiss.

This is how Scoville, the surgeon, and Brenda Milner, the neuropsychologist we encountered in chapter 4, summarized his status in a watershed paper published in 1957:

After operation this young man could no longer recognize the hospital staff nor find his way to the bathroom, and he seemed to recall nothing of the day-to-day events of his hospital life. There was also a partial retrograde [that is, of the past] amnesia, inasmuch as he did not remember the death of a favourite uncle three years previously, nor anything of the period in hospital, yet could recall some trivial events that had occurred just before his admission to the hospital. His early memories were apparently vivid and intact. (Scoville and Milner 1957, 14)

Given the Hippocratic oath of "do no harm," one can only imagine how the surgeon must have felt. In fact, Scoville and Milner stated that one of the goals of their report was to provide a much-needed warning to others about the risk of the procedure. Scientifically, their paper proposed that the hippocampus is "critically concerned in the retention of current experience." In neuroscience, few other reports have spurred such an enormous literature involving both human and animal research. To say that thousands of papers have their origin with their publication is no exaggeration. It might thus come as a surprise to the reader that, six decades later, the exact contributions of the hippocampus to memory remain a matter of intense and heated debate.

In 1971, John O'Keefe and Jonathan Dostrovsky reported that neurons in the hippocampus of the rat respond selectively to places in the environment. Animals were placed in a rectangular box, and when they were in particular locations—say, at the middle of the southmost wall facing south—specific cells in the hippocampus fired vigorously. O'Keefe and Dostrovsky speculated that the region provides the rest of the brain with a "spatial reference map" given that some neurons are particularly attuned to the spatial location, or the place, of the animal in its environment—these neurons would later be popularized as "place cells" (figure 11.3). The implications of these findings were developed extensively in a book by O'Keefe and Lynn Nadel, published in 1978 and now considered a classic—*The Hippocampus as a Cognitive Map*. Since then, intense debate contrasting spatial versus memory functions of the hippocampus has raged in neuroscience.

By treating the hippocampus as a cognitive map, O'Keefe and Nadel attempted to build on the theoretical framework developed in the 1930s by the psychologist Edward Tolman.[6] At a time when learning was viewed as a simple process of passive accumulation of associations imposed on the animal by the environment, Tolman viewed learning as an active process of extracting information from the world. To him, animals track the

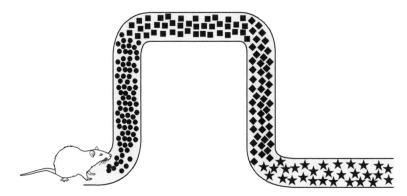

Figure 11.3
"Place cells" in the hippocampus. As the rat navigates down the path, specific neurons fire more vigorously at certain locations. Thus, one neuron fires strongly at the locations marked with stars, another at the locations marked with circles, and so on.

underlying structure of the world through a maplike representation of causal associations (that is, what leads to what?). Its central concept, the cognitive map, allowed the combination of causal information to produce novel ways of achieving outcomes, much like a physical map allows the planning of novel routes to a previously visited destination.

Cognitive maps provide a summary of the places visited by the animal, together with information about distances and directions between them. As an animal moves about its environment, researchers have found that the hippocampus helps encode information about it, including establishing distance and direction vectors: How much distance has it covered and in what directions? Indeed, researchers have by now uncovered various hippocampus neuronal properties that are summarized by catchy descriptors, including "grid cells," "border cells," "head direction cells," "speed cells," and "time cells." (Research on the hippocampus landed John O'Keefe and the Norwegian investigators Edvard Moser and May-Britt Moser the Nobel Prize in 2014; O'Keefe received half the prize and the Moser couple shared the other half.)

As the cell monikers indicate, hippocampal responses are attuned to the spatial and temporal properties during an animal's navigation through its environment. But the more these cells are studied, the clearer it becomes that their activity is very nuanced. Hippocampal firing to places, borders, direction, speed, and so on is influenced by a gamut of factors, including the presence or absence of objects, the presence of a stimulus previously paired with aversive outcomes (as in conditioning paradigms), as well as generally factors such as novelty, attention, and even an animal's internal state (is it hungry?). Other motivational information also plays a role, as cells fire more vigorously near "task goal" locations, including places where an animal receives reward. Together, the firing of hippocampal cells reflects spatial knowledge in extremely rich and multifaceted ways.

Why does the memory-versus-space debate persist to this day? On the surface, it is difficult to appreciate how two such different views of hippocampal function can be reconciled. Is one of them just plain wrong and waiting to be debunked? As often is the case in science, when groups hold opposing views for a long time, both are probably right, at least to some extent. In the present context, one way to square the contrasting views is to think that the brain uses space as a way to organize memories, or what are called episodic memories. Returning to a place, or thinking about it, helps retrieve memories of things and events that happened at that location. A related idea is that the hippocampus generates a "memory map," at once a map of space and a map of memories, together with the links between them.

Despite decades of vigorous work, determining the contributions of the hippocampus to memory remains very much a matter of current scientific interest. However, a noticeable change in today's research is that it is less and less centered on a sole region—it's not all about the hippocampus anymore. Instead, researchers try to understand how the hippocampus interacts with neighboring areas in the temporal lobe (the so-called medial temporal lobe), as well as how it participates in broader interactions with regions across the brain. Episodic memory and spatial navigation are not carried out by single regions—no mental process is.

The Place of Extinction

Extinction is not simply forgetting, or inhibiting, the link between a CS and an aversive event. Considering the environment where the CS and the UCS become uncoupled is a must. In fact, associating environments, rather than just specific cues within them, with safety versus danger is necessary for survival. What anatomical pathways support these processes?

The hippocampus provides context-related information that guides extinction.[7] Animals need to keep track of where extinction took place,

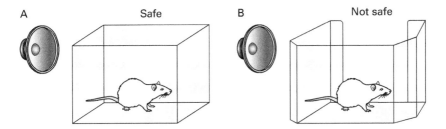

Figure 11.4
Context information is essential for extinction. (a) Original environment context where extinction took place. If the conditioned stimulus (CS) reappears here, the animal is likely safe. (b) If the CS occurs in a new environment, the animal must register this difference and treat it as unsafe. In other words, it is unwise to generalize the safety of the CS across contexts.

because if the CS reappears there, the animal is probably safe. But when the context changes, it makes sense to treat it as dangerous if the CS appears (figure 11.4). The hippocampus has direct connections to the amygdala, and the targeted neurons in the amygdala promote defensive responding ("fear"). Through this pathway, the hippocampus signals that the CS is now happening in a novel context—fear is renewed. The hippocampus also has dense projections to the medial prefrontal cortex, and the hippocampus can engage the medial PFC to indicate that the environment has *not* changed. Here, the original context of extinction is the same experienced presently, so it is likely safe.

Changing Values

Animals learn that some environmental cues are positive signs and predict reward. The pattern of earth around a burrow may indicate to a fox that a mouse has just entered it and that quickly excavating the hideout may lead to catching the prey. After some experience, the animal learns to associate the cue (the earth pattern) with the reward (the mouse). But now suppose that this contingency changes, and the cue is no longer predictive of reward; perhaps mice no longer enter the burrow in a manner that leaves such clear traces. More consequently, the cue may now be associated with a negative outcome. For instance, what if some burrows now contain snakes? In such cases, the animal needs to learn that the cue no longer predicts reward.

Learning to reverse an association, which is known as *reversal learning*, engages the orbitofrontal cortex. A lesion study in the early 1970s demonstrated that monkeys with damage to this area were impaired in their ability to switch or reverse behavior.[8] During reversal learning, the animal first learns that an item is good and predicts subsequent reward, while another item is bad and predicts punishment (or at least is not followed by reward). Training continues until it is clear that the animal has learned the mapping, at which point the association is reversed by the ornery human. Thus, when the first reversed trial is experienced, the animal experiences a complete mismatch between expectation and what is delivered. Without the orbitofrontal cortex, animals had considerable trouble learning the new contingency.

In the early 1980s, investigators managed to record from neurons in the orbitofrontal cortex while monkeys actively performed tasks (Thorpe, Rolls, and Maddison 1983). In some of the first experiments, spiking activity was recorded during reversal learning. Some neurons fired vigorously when the monkey saw a syringe used to deliver black currant juice. But when the contents of the syringe now contained saline (which is mildly aversive, especially in comparison with a favored juice), the monkey's activity declined sharply on seeing the syringe. The discovery of neurons that responded to the *meaning* of the stimulus was quite exciting and led to a wave of experiments trying to sort out the functions of this part of the cortex. In the coming decades, it became increasingly clear that it encodes *value* (figure 11.5): In other words, the firing is not due to particular object features (like the syringe) but to the outcome that it predicts (sweet juice). A corollary of this finding is that distinct objects that signal the same reward generate very similar responses. Reinforcing the notion of value coding, neurons in the orbitofrontal cortex integrate information about the magnitude and the probability of reward. This is critical, because if outcomes are not always certain, one must take into account their probability of occurrence. As one can intuit, a potential reward of $100 that has a probability of 10 percent is not as attractive as an intermediate reward of $50 and a good probability of occurring, say, 70 percent (or even a moderate reward of $25 that is more certain with, say, a 80 percent chance).

Even more broadly, neuronal activity in the orbitofrontal cortex represents the *expectation* of the value of the outcome (Roesch and Schoenbaum 2006). According to this view, the firing of neurons is not simply a result of

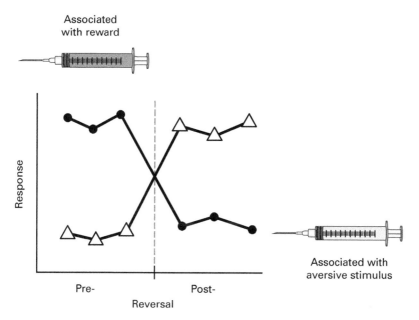

Figure 11.5
Reversal learning and cell responses in the orbitofrontal cortex. Responses to cues predicting "good" (associated with reward) stimuli shift drastically when their meaning is reversed, and vice versa. The vertical line marks when the previously rewarded stimulus now predicts an aversive outcome, and vice versa. The *x* axis indicates blocks of trials before or after reversal.
Source: Results based on Rolls et al. (1996).

the association between an object and its outcome in the past. Instead, it reflects a *prediction* about potential outcomes generated on the fly, at that moment in time. Consider the following "devaluation" experimental paradigm. An animal first learns that different objects go with different food rewards. Say object 1 predicts high reward (a favored food, such as raisins) and object 2 predicts low reward (a less preferred piece of fruit). When the animal sees object 1, neurons in its orbitofrontal cortex fire vigorously, and less so for object 2. When given a choice, the animal will pick object 1 as a means to obtain the favored food. But are the responses to seeing the objects related to the value of the foods? To get at that, experimenters let the animal overfeed on its favored food. What does the monkey do when offered a choice between objects 1 and 2? It will go for object 2 (after all, we've all experienced the decrease in pleasure after overeating sweets!).

Interestingly, monkeys with orbitofrontal cortex lesion do not show a bias for the less favored food; they fail to update the new value of the objects based on their current state (being satiated with the preferred food). They reach out to object 1 even though the food it predicts is not really that desired after consuming so much of it.

We saw how neurons in the hippocampus are particularly sensitive to spatial locations and other navigational information. But cells there are affected by reward and the overall goal relevance of a place, too. Given the involvement of the orbitofrontal cortex in valuation processes, might these two structures work together to integrate spatial information and value? Researchers have started to uncover how interactions between them might be involved. In one experiment, a rat had to navigate its environment in particular ways to receive a reward.[9] In the setup, the animal navigated down an alley and, at certain points, was forced to decide whether to turn right or left (the contraption is called a T-maze given right/left choice points at the top of T-like bifurcations). By trial and error, the rat had to discover which behavior led to a reward—for instance, always turn left at a junction or alternate right and left turns. At the beginning, when rats hadn't figured out the pattern yet, they often paused at the choice point before making a left or right turn. What were they doing? Analysis of both their behavior and cell responses suggests that they were simulating the consequences of potential actions before deciding which turn to take. When they paused just prior to turning, hippocampal cell firing encoded information about pathways ahead of the animal (along the potential left- and right-turn paths), in a manner consistent with trying to determine the consequences of particular actions. Intriguingly, orbitofrontal neurons fired along an entire path based on the probability that the path in question would lead to reward. So, if a segment of the course was part of a pathway leading eventually to reward, firing was vigorous, and vice versa. As the hippocampus has direct connections to the orbitofrontal cortex, the former likely conveys spatial information to the latter so that its behavioral significance can be ascertained—namely, will it lead to a reward?

We've seen that the hippocampus and orbitofrontal cortex encode somewhat overlapping information. Clearly the hippocampus is more attuned to space while the orbitofrontal cortex is more linked to rewards. But instead of thinking of them as implementing different functions—hippocampus:

navigation; orbitofrontal cortex: value—by focusing on their interactions we can see how they support behaviors that are meaningful in natural habitats.

Contrasting Explanations

Let's return to extinction (figure 11.6a). The medial prefrontal cortex in involved in learning that the CS no longer signals threat. However, viewing this region's contribution as the inhibition of emotion by cognition doesn't do justice to the behavior as well as the neuronal interactions at play. One possibility is to conceptualize extinction in terms of the multiple influences discussed so far (figure 11.6b). The final result—extinction—is the result of the multiple contributions, which result in a behavior that is flexible. (We haven't discussed the thalamus in relation to extinction, but based on recent studies, it is one more region that contributes to the process.) Whereas this description broadens the spectrum of influences considerably, it invites thinking that is too linear and region-oriented. The orbitofrontal cortex provides reward information, the hippocampus context, and so on; furthermore, the amygdala responds by combining its inputs to arrive at an answer: Should the animal be wary or not? This type of "boxes and arrows"

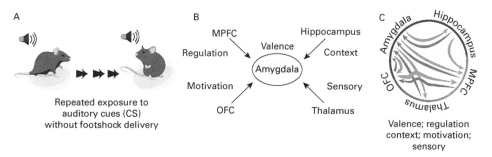

Figure 11.6

Contrasting explanations: (a) Fear extinction. (b) Fear extinction in terms of the top-down regulation of the amygdala by the medial prefrontal cortex, with additional variables influencing the process. (c) Schematic representation of the anatomical connections between some of the brain regions involved, emphasizing a nonhierarchical view of the processes leading to fear extinction. The descriptors "valence," "regulation," and so on are not tied to brain areas in any straightforward one-to-one fashion. MPFC, medial prefrontal cortex; OFC, orbitofrontal cortex.

Source: Reprinted with permission from Pessoa (2018b).

diagram has been used for over a century in neuroscience (Pessoa 2017a), so what could be the alternative?

The diagram in figure 11.6c tries to convey the idea that the brain regions *collectively* determine the extinction process. This type of description provides a different springboard to reasoning about the mapping between function and structure in the brain—how brain regions bring about behaviors. Conceptually, part of the shift is due to the fact that the putative underlying processes (valence, regulation, context, etc.) are *not* separable, so they don't encode stable variables (such as "valence") that are simply pushed up or down by the other factors. Put another way, these variables are so intertwined that they are *jointly determined*: to understand the system, we need to consider the *integration* of the signals.

The scheme diagrammed in figure 11.6c doesn't mean that all regions contribute in the same way to the behavior in question. At first glance, it appears that a lot is missed by diagramming things this way; the description appears too vague. But simplification for the sake of simplification won't help us out of our problem. Go back to figure 11.6a, which says that the behavior in question (exhibiting fear or not) is a function of activity in the amygdala, itself influenced by inputs from the other areas. Figure 11.6b, instead, tells another story: Behavior is a function of all the regions when taken *together*. And their bidirectional interactions imply that the flow of information is far from straightforward (such as the inhibition of the amygdala by the medial PFC).

The boxes-and-arrows arrangement (figure 11.6a) also invites the interpretation that the mechanisms in question are somewhat static. Indeed, in the laboratory, one generally investigates behavior at a specific point in time, or perhaps during a narrow temporal window. But natural behaviors are dynamic, evolving as the animal interacts with its environment. Whereas in the lab an animal may be placed in a chamber where it received shocks in the past, in nature the time frame of experiencing negative (or positive) scenarios is more gradual (unless the animal is surprised by a very sudden attack). Accordingly, a dynamic description of the underlying neural processes is not only beneficial but necessary in general.

Let's try to motivate further the idea of integration of signals and appreciate the nuances of signal flow. As a simple scenario, take a predator-prey system (chapter 8) involving foxes and hare. The number of foxes, F, and the number of hare, H, will fluctuate jointly with time. The number of foxes grows based on predation and decays based on death. We can write this as

$$dF(t) = \alpha FH - \beta F,$$

where dF(t) signifies the change in the number of foxes as a function of time. The first term says that the number grows in proportion to the number of foxes times the number of hare. That is, the increase in foxes is proportional to the number of foxes (the ones that give birth to more of them) and the prey population that supports it. The multiplier α is a constant that specifies the "efficacy" of this growth process (based on predation efficiency and turning food into offspring, for example). The second term says that the number of foxes will decrease from death at a rate given by β.

Now, we need to specify how the hare population changes with time. The number of hare increases exponentially (all that mating!), except that the presence of predators puts that in check:

$$dH(t) = \gamma H - \delta FH,$$

where dH(t) is the change in the number of hare as a function of time. The first term says that the number will grow based on the number of hare present, and the second that it will decrease based on the product of the number of foxes and hare (the more foxes and the more hare, the more encounters and potential death). The constants γ and δ are efficacies of the growth rate (which also accounts for factors such as food availability, and so on) and consumption rate (by foxes), respectively.

These two equations define a "system": To know the number of foxes we need to know the number of hare, and vice versa—they are interdependent. Translating this into the extinction scenario, we can think of the activity of cells in the amygdala, hippocampus, thalamus, medial PFC, and orbitofrontal cortex as jointly interdependent. A computational neuroscientist can then specify equations for how these signals change as a function of time. But how about the experimentalist? How should the experimental scientist proceed? After all, training in neuroscience is not very mathematical. A potential direction is to move research efforts toward studying the *multiregion temporal evolution* of brain data. Here, the focus is on studying multiple regions simultaneously and trying to characterize the joint state of brain regions and how the state evolves temporally (we develop these ideas further in chapter 12). Tools from networks science, among many others, are needed. In the end, experimental scientists need to learn more technical skills or collaborate in larger teams—more likely both.

From Foraging to Escaping

Prey are, by definition, at risk of predation. But their lives aren't always made of the dramatic moments seen in nature documentaries, such as a seal evading a white shark or a gazelle escaping a cheetah, both managing to do so through a series of dazzling twists and turns. Fortunately, they spend a considerable amount of time engaging in positive motivated behaviors, such as foraging, maintaining a nest, nursing, feeding, and mating. They undertake these behaviors when the risk of predation is minimal; obviously, they can't engage in them when they are about to be struck by a predator. But between these two extremes, they exhibit a range of behaviors that depend on their distance to predators. The distance is often the perceived one rather than based on a measuring tape; after all, a predator evades detection exactly to bring its physical separation to the prey within striking distance.

In the late 1980s, Michael Fanselow and Laurie Lester proposed that prey behaviors are structured around a *continuum of predatory imminence* with particular key stages: pre-encounter, post-encounter, and circa-strike.[10] In the absence of predators, animals will engage in their preferred activities, including the positive behaviors mentioned above. How they behave the rest of the time takes into account predator distance. During pre-encounter, behavior is based on the assessment of the probability of encountering a predator. So, although a predator hasn't been detected, foraging may proceed more cautiously in places where predators have been spotted in the past. During post-encounter, the animal's behavior shifts quite strikingly; they may suppress behavior, taking stock of the situation (should they dash away or can they continue grazing for a little longer?). Circa-strike behaviors are often a last-ditch attempt to escape from capture and are often unusual and highly energy consuming. For example, a deer mouse freezes in the presence of a gopher snake but attempts a last second, spectacular vertical leap as the snake strikes.

Investigators originally described these stages based on stereotypical and relatively fixed patterns; yet actual behaviors are quite flexible. One can think more broadly in terms of computations of *threat detection* and *threat escape.*[11] The central goal of threat detection is to evaluate sensory information to determine if a threat is present. This assessment of threat is flexible and dynamic and calibrated by expectations built on experience. If the risk of predation is low, prey adjust the threshold for reacting to threats to a higher level (more evidence is needed) compared to when the risk is higher (less evidence is needed). Remarkably, animals quickly learn to suppress escape responses if they are repeatedly challenged but no adverse outcome ensues, even if the stimuli are potent and innately threating. In all, the choice of action when threat is spotted is highly context dependent. Threat detection has not been studied extensively in mammals, and little is known about the underlying mechanisms (more research has been conducted in invertebrates, as well as fishes, frogs, and birds). But we know that the superior colliculus (chapter 3) participates in the detection process. For example, visual signals from the retina engage the neurons in the superior layers of this structure, which are tuned to looming stimuli resembling a predator coming from above.

The ability to handle threats expands considerably with learning, and in particular, animals learn to avoid locations where predators were encountered previously. For example, when exploring an arena where they saw threats before, mice attempt to escape. (Most of the focus in rodent research has been on freezing responses instead of active escaping because the chambers used are small and don't provide a possible escape route.) Alternatively, mice generate other defensive behaviors, such as an increase in stretch postures or reduced exploratory locomotion, which indicates their altered risk assessment of the situation (Silva et al. 2013).

Naturally, after a stimulus is detected and deemed a threat, an action is required, and fast—this is the escape computation referred to before. Indeed, evolution has shaped some neural circuits to permit just that. Specialized cells in fish, for example, allow responses to start a mere 5 to 10 milliseconds after threat is surmised (this fast response allows fish to change direction and be propelled forward). But animals do not necessarily flee immediately once predators are detected. Why? In a nutshell, running away is not always a good idea. Field studies reveal that animals attempt to get away when the costs of remaining (such as the risk of injury or capture) are higher than the cost of fleeing (such as loss of foraging opportunities). There's also a close link between an animal's internal state and its decision to escape. For example, mice exhibit risky behaviors when hungry, such as spending more time in threatening environments. Sexual receptiveness also influences the escape calculus.

A frequent alternative to fleeing is, as discussed in chapter 3, freezing in place. The choice between staying or going is determined, in part, by knowledge about the spatial properties of the environment. For example, mice memorize an escape location based on a single and brief (less than 20 seconds) visit to a shelter, and changes to the spatial environment lead to a rapid update of the defensive action chosen: fleeing versus freezing. Additional variables considered by mice include how safe the shelter is, the distance and relative position of the predator, and potential competition for shelter access.

Animals thus confront a *detection-response dilemma*: both responding too early and too late are costly. Escaping is metabolically expensive as getting away requires energy. But it is also costly in terms of opportunity losses, including those related to food and mating. In effect, triggering a full-blown escape response on detecting a threat is a rather poor survival strategy and essentially nonexistent in the natural kingdom. The decision to take flight is not just triggered by threat detection and involves computations that rely on multiple external and internal variables. Together, escape behaviors are far from simple stimulus-driven, stereotypical reactions. The mechanisms involved engage specialized circuits refined by eons of evolutionary time. Whereas some circuit components play special roles, they exchange information with multiple areas across the brain. The behavioral complexity points to solutions involving the integration of signals across distributed circuits so as to promote behavioral flexibility and survival.

As mentioned, little is known about the brain circuits involved in escape in mammals. What we do know has focused on a few usual suspects: the superior colliculus, periaqueductal gray (PAG), hypothalamus, and amygdala. To understand why knowledge is so limited, we need to consider the inherent limitations of current experimental setups. A typical rodent experiment will take place in a small cage, where the animal is exposed in a controlled fashion to stimuli such as tones, lights, or a foot shock. The animal will also have levers to press and simple decisions to make (choose food A versus B). If brain recordings are being made, the animal is tethered so that electrode signals can be conveyed to a computer for data analysis. To be sure, behavioral experiments without recordings are performed in larger setups, including T- or radial-shaped mazes. But even in these cases the restrictions are considerable. And the lesion method, a valuable but coarse instrument, has been the mainstay of investigators. But neuroscience is changing fast. Large environments are being used more frequently; experiments with untethered animals are becoming more prevalent; and genetic and chemical manipulations allow investigators to focus on subclasses of cells in specific brain regions, such as a particular population of cells in the basolateral amygdala that has specific chemical properties. Exciting days lie ahead of us.

The overall goal of this chapter is to illustrate how an ostensibly simple behavior—fear extinction—is implemented in the brain. Far from simple, extinction is a complex learning process supported by distributed brain circuitry. Although we barely scratched the surface, we saw that only by acknowledging the simultaneous contributions of several brain areas can we hope to do justice to the intelligence of the behavior. What's more, the interdependence of the regions leads to a process of joint construction of a solution by the brain—a solution with many authors, each of which contributes different materials.

12   It's All about Complex, Entangled Networks

We've come to the end of our short exploration of the brain, this most mysterious of organs. This book has tried to illustrate how mental processes are built from intricate interactions involving gray and white matter components. We've learned, hopefully, to appreciate some of the complexity of how the brain contributes to bringing forth the mind. In this last chapter, we return to some of the big questions and problems encountered previously, and some of the themes that will be important for advancing our understanding of mind and brain in the future.

What Does Brain Evolution Mean?

The geneticist Theodosius Dobzhansky famously stated that in biology, nothing makes sense unless it's in light of evolution. The same applies to neuroscience, a biological science. But evolution poses a conundrum. Vertebrates have been evolving for over 500 million years. A telencephalon, a midbrain, and a hindbrain are part of the general plan of their nervous system. Structures like the amygdala and the striatum are found in animals as diverse as a salmon, a crow, and a baboon. Thus, many parts of the brain are "conserved." But then, what is novel? Something *must* be new, after all.

In chapter 9, we reviewed how *homology* refers to relationships between traits shared as a result of common ancestry. The leaves of plants provide a good example.[1] The leaves of a pitcher plant, Venus flytrap, poinsettia, and a cactus look nothing alike and, in fact, have distinct functions. In the pitcher plant, the leaves are modified into pitchers to catch insects; in the Venus flytrap, they turn into jaws to catch insects; a poinsettia's bright red leaves resemble flower petals and attract insects and pollinators; leaves on a cactus plant have become modified into spines, which reduce water loss and can protect the plants from herbivores. Nothing alike—yet the four are homologous because they derive from a common ancestor.

A structure adopts new functions during evolution, while its ancestry can be traced to something more fundamental.[2] Take the hippocampus of rodents, monkeys, and humans. There is copious evidence indicating that the area is homologous in the three species—that is, it's a conserved structure. But does it perform the same function(s) in these species, or does it carry out *qualitatively* different function(s) in humans, for example? To many neuroscientists, this sounds implausible. However, the possibility need not be any more radical than saying that the forelimb does something qualitatively different in birds compared to turtles, say. If common ancestry precluded new functions, no species could ever take flight!

The ongoing discussion is particularly pertinent when we think of emotion and motivation, because researchers invoke "old" structures when studying these mental phenomena. Regions like the amygdala at the base of the forebrain and the periaqueductal gray (PAG) in the midbrain are invoked in the case of emotion, and the accumbens (part of the striatum) also at the base of the forebrain and the ventral tegmental area in the midbrain in the case of motivation. Because these regions are deeply conserved across vertebrates, they function in a similar way, or so the reasoning goes. If we entertain these areas in rodents, monkeys, and humans, closer as they are evolutionarily, the expectation would be that they work similarly. But rodents and primates diverged more than 70 million years ago. Are we to suppose that no *qualitative* differences have emerged? This seems rather implausible. (In chapter 9, we briefly reviewed some structural differences in the amygdala of rats, monkeys, and humans.)

The argument made in this book is that we should conceptualize evolution in terms of the *reorganization* of larger-scale connectional systems. Instead of more cortex sitting atop the subcortex in primates relative to rodents—which presumably allows the "rational" cortex to control "primitive" parts of the brain—more varied ways of interactions are possible, supporting more mental latitude.

The brain doesn't fossilize. Unfortunately, with time, it disintegrates, leaving no trace. So we simply don't have a way to know exactly what the brain of a common ancestor looked like. Without fossil remains, scientists tend to think of the brain of a common ancestor of rodents, primates, and humans as something like the current brain of a mouse, as this animal is the "rudimentary" one. But a mouse encountered today has had 75 million years to evolve from the ancestor in question, ample time to specialize to the particular niches it inhabits now.

Evolution is as much about what's preserved as what's new. Ever since science was transformed by the independent work of Charles Darwin and Alfred Russel Wallace in the late 1850s, biologists have sought to determine "uniquely human" characteristics. This has led to a near-obsession to identify one-of-a-kind nervous system features, from putative exclusively human brain regions to cell types. The cortex, in particular, has attracted much attention. The pallium of mammals is structured in a layered fashion, a quality not observed in other vertebrates. Well, not exactly, as some reptiles (such as turtles) have a dorsal pallium that is cortex-like, with three bands of cells. Mammals, however, have parts of the cortex that are much more finely layered, with six well-defined zones. In fact, a six-layered cortex is often referred to as "neocortex," with the "neo" part highlighting its sui generis property (in the book, the more neutral terminology "isocortex" was used in chapter 9 for this type of cortex).

I believe that the concept of reorganization of circuits is a much more promising idea. That is to say, what is unique about humans is the same that is unique about mice, or any other species: Their circuits are wired in ways that support survival of the species. This is not to deny that some more punctate differences play a role. But whatever the differences are, at least considering primates with larger body sizes, they are not staring us in the face—they are subtle. For example, all primates exhibit an isocortex that is massively expanded. Primates also have prefrontal cortices with multiple parts, including the lateral component, which neuroscientists often link to "higher cognitive" capabilities. More generally, direct evidence for human-specific cortical areas is scant.

Let's go back to Dobzhansky's call to consider biology in light of evolution— always. Biologists would vehemently agree. But evolution is so egregiously complex that the suggestion doesn't help as much as one would think. What we observe in practice is that neuroscientists who don't specialize in studying brain evolution are time and again cavalier, if not outright naive, about how they apply and think of evolution. By doing so, our explanations run the risk of becoming *just-so stories.*[3]

Fitting Behavior Inside a 40 × 40 × 40 Centimeter Box

The central question in neuroscience is to understand the physical basis of behavior. But what kinds of behavior can be studied in a lab? Mice and rats can be placed in chambers and mazes to perform tasks. One can then study the effects of lesions on behavior. If cell recordings are conducted, the constraints are even more severe. Until just a few years ago, this required a fair amount of cabling to link the brain to signal amplifiers and other electronics. Experiments in primates are performed in a "monkey chair" that keeps the animal's body and head in place. Humans, of course, are studied inside magnetic resonance imaging (MRI) tubes that are anything but organic. With the technology available, getting closer to natural behaviors has simply not been possible.

A type of behavior that fits inside a $40 \times 40 \times 40$ centimeter box is classical conditioning. Indeed, it has been extensively studied by psychologists since the early twentieth century, and for those interested in the biological mechanisms of fear, the paradigm has been a godsend. It offers a window into this process while allowing careful control over study variables, a fundamental consideration in experimental science. The neuroscience of fear has been one of the most active areas of inquiry, thanks to the paradigm.

But the fixation with this task has led to a form of tunnel vision.[4] As Dennis Paré and Gregory Quirk, very prominent researchers in this area, state:

> When a rat is presented with only one threatening stimulus in a testing box that allows for a single reflexive behavioral response, one is bound to find exactly what the experimental situation allows: neuronal responses that appear tightly linked to the CS [conditioned stimulus] and seem to obligatorily elicit the conditioned behavior. (Paré and Quirk 2017, 6)

The very success of the approach has led to shortsightedness.

Placed inside a small, enclosed chamber the animal is limited to a sole response: Upon detecting the CS, it ceases all overt behavior and freezes in place. It can't consider other options, such as dashing to a corner to escape; it cannot try to attack the source of threat either, as there isn't another animal around—the shock comes out of nowhere! Now, when researchers study the rat's brain under such conditions, a close relationship between brain and behavior is established. But as Paré and Quirk warn, the tight link might be apparent insofar as it would not hold under more general conditions.

Neuroscience is experiencing a methodological renaissance. Advances in chemistry and genetics now allow precision in targeting regions and circuits in ways that would have sounded like science fiction a decade ago. But if we continue using the paradigms that have been the mainstay of the field, we will be cornering ourselves into a scientific cul-de-sac.[5] It's time to think outside the box.

A Thought Experiment

Try to contemplate a future device that allows registering in minute detail the behaviors of a cheetah and a gazelle during a chase, including all muscle and skeletal movements. At the same time, we are capable of recording billions of neurons across the two nervous systems while the entire chase unfolds from before the cheetah initiates the pursuit until its dramatic conclusion. What would we discover? How much of our textbooks would have to be altered?

A radical rethinking might be needed, and a lot would have to be rewritten. Alternatively, many of the experimental paradigms employed to date are quite effective in isolating critical mechanisms that reflect the brain's functioning in general settings. True, novel findings would be made with new devices and techniques, but they would extend current neuroscience by building naturally on current knowledge. The first scenario is not idle speculation, however.

So-called naturalistic experimental paradigms are starting to paint a different picture of amygdala function, for example. In one study, a rat was placed at one end of an elongated enclosure and a piece of food placed midway between the rat and a potential predator, a Lego-plus-motor device called a "Robogator" (figure 12.1) (Amir et al. 2019). To successfully obtain the food pellet, the rat had to retrieve it before being caught by the Robogator (don't worry, capture didn't occur in practice). The findings were inconsistent with the standard "threat-coding model" that says that amygdala responses reflect fear-like or other related defensive states. During foraging, when the rats were approaching the pellet, neurons reduced their firing rate and were nearly silent, not active, close to the predator. Clearly, responses did not reflect a threat per se.

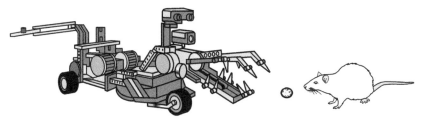

Figure 12.1
More naturalistic experimental paradigms are starting to be employed. Here, a rat can acquire a piece of food but must catch it before being chased by the "Robogator." *Source*: Model for the "Robogator" at left kindly provided by Jeansok J. Kim.

Another study recorded from neurons in the amygdala over multiple days as mice were exposed to different conditions (Gründemann et al. 2019). The mice were exposed to a small open field and were free to explore it. These creatures don't like to feel exposed, so they stayed at the corners of the box a good amount of time. But they also ventured out into the open and walked around the center of the box with some frequency. The researchers discovered two groups of cells: one engaged when the mouse was being more defensive in the corners (these "corner" cells fired vigorously at these locations), another when the mouse was in an exploratory mode visiting the center of the space ("center" cells fired strongly when the animal was around the middle of the field). The researchers recorded from the exact same cells during more standard paradigms, too, including fear conditioning and extinction. They then tested the idea that the firing of amygdala neurons tracks "global anxiety." For instance, they should increase their responses when the animal entered the center of the field in the open-field condition, as well as when they heard the CS tone used in the conditioning part of the experiment. Surprisingly, cells did not respond in this way. Instead, neuronal firing reflected moment-to-moment changes in the exploratory state of the animal, such as during the time window when the animal transitioned from exploratory (for example, navigating in the open field) to nonexploratory behaviors (for example, when starting to freeze).

The above two examples provide tantalizing inklings that there is a lot to discover—and revise—about the brain. It's too early to tell, but given the technological advances neuroscience is witnessing, examples are popping up all over the place. For example, a study by Karl Deisseroth and colleagues recorded activity of approximately 24,000 neurons throughout 34 brain regions (cortical and subcortical).[6] Whereas measuring electrical activity with implanted electrodes typically measures a few cells at a time, or maybe about 100 by using state-of-the-art electrode grids, the study capitalized on new techniques that record calcium fluorescence instead. When cells change their activity, including when they spike, they rely on calcium-dependent mechanisms. In genetically engineered mice, neurons literally glow based on their calcium concentration. By building specialized microscopes, it is possible to detect neuronal signaling across small patches of gray matter. In their study, when mice smelled a "go" stimulus, a licking response produced water as a reward. The animals were highly motivated to perform this simple task as the experimenters kept them in a water-restricted state. Water-predicting sensory stimuli (the "go" odor) elicited activity that rapidly spread throughout the brain of the thirsty animals. The wave of activity began in olfactory regions and was disseminated within approximately 300 milliseconds to neurons in every one of the 34 regions they recorded from! Such propagation of information triggered by the "go" stimulus was not detected in animals allowed to freely consume water. Thus, the initial water-predicting stimulus initiates a cascade of firing throughout the brain only when the animal is in the right state—thirsty.

In another breakthrough study, researchers used calcium imaging to record from more than 10,000 neurons in the visual cortex of the mouse, while facial movements were filmed in minute detail (Stringer et al. 2019). They found that information in cortical neurons reflected over a dozen features of motor information (related to facial movements, including whiskers and other facial features), in line with emerging evidence from other investigations. These results are remarkable because, according to traditional thinking, motor and visual signals are only merged later in so-called higher-order cortical areas and definitely not in the primary visual cortex. But the surprises didn't stop there. The researchers also recorded signals across other parts of the forebrain, including cortical and subcortical areas. Surprisingly, information about the animal's behavior (at least as conveyed by motor actions visible on the mouse's face) was observed nearly everywhere they recorded. In considering the benefits of such ubiquitous mixing of sensory and motor information, the investigators ventured that behaving effectively depends on the combination of sensory data, ongoing motor actions, and internal-state variables. It seems that this is happening pretty much everywhere, including in parts of the brain believed for a long time to not mix them, like the primary sensory cortex.

The examples above hint that much is to change in neuroscience in the coming decades. Still, these results come from fairly constrained experimental settings. The amygdala study used a $40 \times 40 \times 40$ centimeter plastic box; the thirst study probed mice with their heads fixed in placed; and the facial movement study employed an "air-floating ball" that allowed mice to "run." Imagine what we'll discover in the future!

How Complicated Is It Again?

Throughout the book, we've been describing a systems view that encourages reasoning about the brain in terms of large-scale distributed and entangled circuits. But when we adopt this stance, pretty early on it becomes clear that "things are complicated." Indeed, we might be asked if we need to entangle things that much. Shouldn't we attempt simpler approaches and basic explanations first? After all, an important principle in science is *parsimony*, often discussed in reference to what's called Occam's razor after the Franciscan friar William of Ockham's dictum that *pluralitas non est ponenda sine necessitate*: "Plurality should not be posited without necessity." In other words, keep it simple, or as Einstein is often quoted, "Everything should be made as simple as possible, but no simpler."

This idea makes sense, of course. Consider a theory T that tries to explain a given set of phenomena. Now suppose that an exception to T is described—say, a new experimental observation that is inconsistent with it. While not good for proponents of T, the finding need not be the theory's death knell if it's possible to extend T so that it can handle the exception, thereby avoiding the theory from being falsified. As T breaks down further and further, it could be gradually extended to explain the additional observations with a series of, possibly, ad hoc extensions. That's clearly undesirable. At some point, the theory in question is so bloated that simpler explanations would be heavily favored in comparison.

But whereas parsimony is abundantly reasonable as a general approach, what counts as "parsimonious" isn't exactly clear. That's where the rubber hits the road. Take an example from physics. In 1978, an American astronomer, Vera Rubin, noticed that stars in the outskirts of galaxies were rotating too fast, contradicting what would be predicted by our theory of gravity.[7] It was as if the mass observed in the universe was not enough to keep the galaxies in check, triggering a vigorous search for unaccounted sources of mass. Perhaps the mass of black holes had not been tallied properly? Other heavy objects such as neutron stars? When everything known was added up, the discrepancy was—and still is—huge. Actually, about five times more mass would be needed than what we have been able to put on the scale.

To explain the puzzle, physicists postulated the concept of dark matter, a substance previously unknown and possibly made of as-yet undiscovered subatomic particles. This would account for a whopping 85 percent of the mass of the universe. We see that to solve the problem, physicists left the standard theory of gravitation (let's call it G) unmodified, but they had to postulate an entirely new type of matter (call it m). This not a small modification! In 1983, the Israeli physicist Mordehai Milgrom proposed an alternative solution, a relatively small change to G. In his "modified gravity" theory, this force works as usual except when considering such massive systems as entire galactic systems. Without getting into the details, the change essentially involved adding a new constant, called a_0, to the standard theory of gravity.[8]

So, our old friend G doesn't work. We can either consider {G + m} or {G + a_0} as potential solutions. Physicists have not been kind to the latter solution and considered it rather ad hoc. In contrast, they have embraced the former and devoted monumental efforts to finding new kinds of matter that can tip the scales in the right direction. At present the mystery is unsolved, and larger and more sophisticated instruments continue to be developed in the hope of cracking the problem. The point of this brief incursion into physics was not to delve into the details of the dispute but to illustrate that parsimony is easier said than done. What is considered frugal in theoretical terms depends very much on the intellectual mindset of a community of scientists. And, human as they are, they disagree.

Another example that speaks to parsimony relates to autonomic brain functions that keep the body alive—for example, regulating food and liquid intake, respiration, heart activity, and the like. The anatomical interconnectivity of this system has posed major challenges to deciphering how particular functions are implemented. One possibility, along the lines proposed in this book, is that multiregion interactions collectively determine how autonomic processes work. But consider an alternative position. In an influential review, Clifford Saper stated that "although network properties of a system are a convenient explanation for complex responses, they tell us little about how they actually work, and the concept tends to stifle exploration for more parsimonious explanations" (Saper 2002, 460). According to him, the "highly interconnected nature of the central autonomic control system has for many years served as an impediment to assigning responsibility for specific autonomic patterns" to particular groups of neurons.

It's not a stretch to say that thinking in terms of complex systems is not entirely natural to most biologists. Many, in fact, view it with a nontrivial amount of suspicion, as if this approach "overcomplicates" things. Their training emphasizes other skills, after all. Unfortunately, if left unchecked, the drive toward simple explanations can lead researchers to adopt distorted views of biological phenomena, as when proposing that a "schizophrenia gene" explains this devastating condition or that a "social cognition brain area" allows humans, and possibly other primates, to have behavioral capabilities not seen in other animals. Fortunately, neuroscience is gradually changing to reflect a more interactionist view of the brain. From the earlier goal of studying how regions work, current research takes to heart the challenge of deciphering how circuits work.

A key aspect of any scientific enterprise is conceptual. Scientists decide the important questions that should be studied by accepting or rejecting papers in the top journals, funding particular research projects, selecting topics for conferences, and so on. Many of these judgments are subjective, in the sense that they are not inherent to the data collected by scientists. How one studies natural phenomena is based on accepted approaches and methods of practicing researchers. Accordingly, the position to embrace or shun complex systems is a collective viewpoint. To some, network-based explanations are too unwieldy and lacking in parsimony. In diametrical contrast, explanations heavily focused on localized circuits can be deemed as oversimplistic reductionism, or purely naive.

In the end, science is driven by data, and the evidence available puts pressure on the approaches adopted. Whereas mapping the human genome took $3 billion and a decade at first, gene mapping can be done routinely now for under a thousand dollars in less than two days, enabling unprecedented views of genetics. In the case of the brain, it's now viable to record thousands of neurons simultaneously, opening a window into how large groups of neurons generate behaviors in ways that weren't possible before. (Remember that most of what we know about neurophysiology relied on recordings of individual neurons or very small sets of cells at a time.) For example, in a trailblazing study, researchers recorded single neurons across the entire brain of a small zebrafish.[9] Their goal was to record not most but *all* of the creature's neurons! Perhaps one day in the not-so-distant future, the same will be possible for larger animals.

Causation in Complex Systems Is a Whole Different Thing

Nowhere else is the challenge of embracing complex systems greater than when confronting the problem of *causation*. "What causes what" is *the* central problem in science, at the very core of the scientific enterprise.

One of the missions of neuroscience is to uncover the nature of signals in different parts of the brain and ultimately what causes them. A type of reasoning that is prevalent is what I've called the *billiard ball* model of causation (Pessoa 2017a, 2018b). In this Newtonian scheme, force applied to a ball leads to its movement on the table until it hits the target ball. The reason the target ball moves is obvious; the first ball hits it, and through the force applied to it, it moves. Translated into neural jargon, we can rephrase it as follows: A signal external to a brain region excites neurons, which excite or inhibit neurons in a second brain region through anatomical pathways connecting them. But this way of thinking, which has been very productive in the history of science, is too impoverished when complex systems—the brain for one—are considered.

We can highlight two properties of the brain that immediately pose problems for standard, Newtonian causation.[10] First, anatomical connections are frequently bidirectional, so physiological influences go both ways, from A to B and back. If one element causally influences another while the second simultaneously causally influences the first, the basic concept breaks down. Situations like this have prompted philosophers to invoke the idea of "mutual causality." For example, consider two boards arranged in a Λ shape so that their tops are leaning against each other; so, each board is holding the other one up. Second, *convergence* of anatomical projections implies that multiple regions concurrently influence a single receiving node, making the attribution of unitary causal influences precarious.

If the two properties above already present problems, what are we to make of the extensive cortical-subcortical anatomical connectional systems and, indeed, the massive combinatorial anatomical connectivity discussed in chapter 9? If, as advanced, the brain basis of behavior involves distributed, large-scale cortical-subcortical networks, new ways of thinking about causation are called for. The upshot is that Newtonian causality provides an extremely poor candidate for explanation in *non-isolable* systems like the brain.

What are other ways of thinking about systems? To move away from individual entities (like billiard balls), we can consider the temporal evolution of "multiparticle systems," such as the motion of celestial bodies in a gravitational field. Physicists and mathematicians have studied this problem for centuries, which was central in Newtonian physics. For example, what types of trajectories do two bodies, such as the earth and the sun, exhibit? This so-called two-body problem was solved by Johann Bernoulli in 1734. But what if we are interested in three bodies—say, we add the moon to the mix? The answer will be surprising to readers who think the challenge sounds easy given that two bodies were understood long ago. On the contrary, this problem has vexed mathematicians for centuries and in fact cannot be solved! At least not in the sense of two bodies, because it doesn't admit to a general mathematical solution.

So, what can be done? Instead of analytically solving the problem, one can employ the laws of motion based on gravity and use computer *simulations* to determine future paths.[11] If we know the position of three planets at a given time, we can try to determine their positions in the near future by applying equations that explicitly calculate all intermediate positions. In the case of the brain, where we don't have comparable equations, we can't do the same. But we can extract a useful lesson and think of the joint state of multiple parts of the brain at a given time. How does this state, which can be summarized by the activity level of brain regions, change with time?

Before describing some of these ideas further, I'll propose an additional reason that contemplating dynamics is useful. For that, we need to go back in time a little.

The World Is Made of Processes, Not Things

Starting in the mid-fifth century BCE, Greek thinkers including Leucippus, Democritus, and Epicurus thought of nature as made of immutable atoms in empty space. Although the scientific revolution of the sixteenth and seventeenth centuries shunned older classic ideas, it enshrined atomism. That is to say, first and foremost the world consists of substantial particles or *things*. Science, therefore, must seek to explain organs, cells, molecules, and so on, to give a few biological examples. This statement appears so innocuous and obvious as to appear to be a truism. What could possibly be an alternative?

Perhaps not surprisingly, we can go back to the Greeks to entertain a second view, one encapsulated in the dictum *panta rhei* ("everything flows"), or the famous saying that "no person ever steps in the same river twice." As Heraclitus suggested:[12]

> Reality is not a constellation of things at all, but one of processes. The fundamental "stuff" of the world is not material substance, but volatile flux, namely "fire", and all things are versions thereof (puros tropai). Process is fundamental: the river is not an object, but a continuing flow; the sun is not a thing, but an enduring fire. Everything is a matter of process, of activity, of change (panta rhei).

In the first half of the twentieth century, this view was embraced by the so-called organicists, biologists attempting to build a science of life that was dynamic and systems oriented. As asserted by the geneticist Conrad

Waddington, biology does not study things; it studies processes occurring at various timescales (Dupré and Nicholson 2018, 9). In this view, thinglike entities don't necessarily need to be excluded; they can be considered "processes" stable and sustained enough to have substance—think of a person, an organ, or a cell. But embracing such a process-oriented mindset—a *process philosophy*—naturally leads to new ways of formulating and answering scientific questions. And whereas this view obviously helps with phenomena such as hurricanes, streams, and vortices, it encourages describing biological phenomena in terms of context-dependent, dynamic processes. For one, the compulsion of neuroscientists to define areas and subareas recedes, giving way to the goal of deciphering how processes involving multiple parts of the brain unfold temporally to support behaviors.

Transient Brain Dynamics

If the discussion sounds intriguing, it almost certainly feels vague. Let me illustrate some ideas in use by neuroscientists to understand *multiregion dynamics*. The objective is to describe the *joint state* of a set of brain regions and how it changes.

Imagine a system of n brain regions labeled 1, 2, . . . , each with an activation (or firing rate) strength that varies as a function of time denoted $x_1(t)$, $x_2(t)$, and so on. We can group these activities into a vector x. Recall that a vector is simply an ordered set of values, such as x, y, and z in three dimensions. At time t_1, the vector $x(t_1)$ specifies the *state* of the regions (that is, their activations) at time t_1. By plotting how this vector moves as a function of time, it is possible to visualize the temporal evolution of the system. We can call the succession of states at t_1, t_2, etc., visited by the system a *trajectory*. Now, suppose an animal performs two tasks, A and B, and that we collect responses across three brain regions, at multiple time points. We can then generate a trajectory for each task (figure 12.2), each providing a potentially unique *signature* for the task in question.[13] We have created a four-dimensional representation of each task, and it considers three locations in space (the regions where the signals were recorded from) and one dimension of time. Of course, we can record from more than three places; that only depends on what our measuring technique allows us to do. If we record from n spatial locations, then we'll be dealing with an $(n+1)$-dimensional situation (the +1 comes from adding the dimension of time). Whereas we can't plot

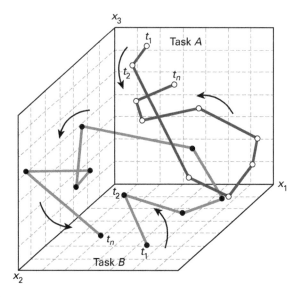

Figure 12.2
Spatiotemporal trajectories. The activity of three brain areas is shown across time. The three values at time t form a vector. In three dimensions, we can plot vectors (see figure 4.3b for examples) at each time point and join the endpoints. In this representation, time is only implicit and runs along the lines starting at t_1 and ending at t_n. The activities across the three regions as a function of time evolve distinctly for task A and task B. This particular evolution, the trajectory, provides a signature for the task in question.

the trajectory on a piece of paper, fortunately the mathematics is the same, so it poses no problems for data analysis.

When we think in terms of spatiotemporal trajectories, the object of interest—the trajectory—is spatially distributed and, of course, dynamic. It also encourages a process-oriented mindset instead of trying to figure out how a brain region responds to a brief stimulus. The process view also changes the typical focus on "billiard ball" causation—the white ball hits the black ball, or region A excites region B—as we are less obsessed about what single factor is responsible for a region's response. Experimentally, a central goal, then, becomes estimating trajectories robustly from available data.

Some readers may feel that, yes, trajectories are fine, but aren't we merely describing the system but not *explaining* it? Why is the trajectory of task A different from that of task B, for example? Without a doubt, a trajectory is not the be-all and end-all of the story. Deciphering how it comes about is ultimately the goal, which will require more elaborate explanations, and here computational models of brain function will be key. In other words, what kind of system, and what kind of interactions among system elements generate similar trajectories, given similar inputs and conditions?

Final Thoughts

Neuroscience strives to elucidate the neural underpinnings of behaviors. Modern neuroscience has done so in a preponderantly reductionistic fashion for over a century and a half.[14] I would venture that progress has been stymied by such approach and that the time is ripe for the field to phase-transition into a period when a truly dynamic and networked view of the brain takes hold. Future research will need to strive to make progress along several fronts: dynamics, decentralized computation, emergence, and competition. At the same time, a science of the mind-brain must be developed by erecting it from a solid foundation of understanding behavior while employing computational and mathematical tools in an integral manner.

I believe the field of neuroscience needs to take stock and invest on the development of conceptual and theoretical sides. Bigger and shinier tools and techniques alone won't yield the necessary progress; we run the risk of being able to measure every cell (or subcellular component even) in the brain in a theoretical vacuum. To drive the point home, suppose that experimental physicists could measure every atom of a given galaxy. How would that advance understanding if not for a theory of gravitation that took more than 400 years of development? The current obsession in the field with causation is equally problematic. Without conceptual clarity (how should we even think of causation in highly entangled systems?), "causal" explanations in fact might miss the point.

Ultimately, to explain the cognitive-emotional brain, we need to dissolve boundaries within the brain[15]—perception, cognition, action, emotion, motivation—as well as outside the brain, as we bring down the walls between biology, psychology, ecology, mathematics, computer science, philosophy, and so on. Only then we will be on the right track.

Glossary

Accumbens: Part of the **ventral striatum** that receives a large concentration of dopamine-carrying axons from the midbrain; also called the *nucleus accumbens*.

Action potential: An electrical signal that propagates along the neuron's extension (see **axon**), much like current flowing along a wire.

Amygdala: Subcortical structure extensively studied in the context of aversive conditioning but involved in a very large array of functions. The "amygdala complex" contains at least a dozen subnuclei. In very broad terms, it is useful to consider a "basolateral amygdala" and a "central amygdala."

Area: A unit that is thought to be anatomically and functionally meaningful by many neuroscientists. In the book, an area is viewed as functionally meaningful when it participates in large-scale, distributed circuits. In the cortex, an area is typically defined based on cytoarchitectonic differences of lamination pattern, including cell type and density (example: primary visual cortex in occipital lobe). In the subcortex, an area is typically defined based on patterns of cell type and density, as well as other markers of putative boundaries (example: **amygdala**; see figure 5.5).

Autonomic nervous system: Part of the peripheral nervous system that is connected with glands and smooth muscle lining organs (such as the heart). Thus, it influences the function of internal organs, which influence the central nervous system in turn.

Axon: The extension of the neuron that comes into close contact with other neurons and typically affects the dendrite of a postsynaptic cell via the synapse.

Basal ganglia: Typically refers to a group of subcortical nuclei, including the **striatum** (caudate plus putamen) and the globus pallidus.

Brainstem: The posterior stalklike part of the human brain that connects the **cerebrum** with the spinal cord; it is composed of the midbrain, the pons, and the medulla oblongata.

Central nervous system: Consists primarily of the brain and the spinal cord. In contrast, the peripheral nervous system consists of the nerves and ganglia (that is, masses of cells) outside the brain and spinal cord.

Cerebrum: The largest part of the brain containing the cerebral cortex as well as several subcortical structures, including the hippocampus, basal ganglia, and olfactory bulb; also called the **telencephalon**.

Complex system: One in which the properties of the system are highly dependent on interactions of its many parts, possibly involving feedback and cycles, as well as nonlinearities. In this context, "complex" should not be equated with "complicated," which would refer to a system that has many components.

Conditioned stimulus (CS): An initially neutral stimulus that gains affective significance by being paired with an **unconditioned stimulus (UCS)**, such as one that is inherently aversive or negative (such as a shock). A conditioned stimulus can also acquire positive significance by being paired with a unconditioned rewarding stimulus or event (such as food).

Cortex: The cerebral cortex, also known as the cerebral mantle, is the outer gray matter neural tissue of the cerebrum of the brain in mammals. It is organized into a series of layers.

Dendrite: The extension of the neuron that typically receives stimulation (from axons).

Emergence: Idea that system properties arise from the nonlinear combination of its parts. Thus, characterizing the parts of a system and combining them by "adding them up" misses important system properties.

Extinction: Learning process by which a **conditioned stimulus** (see definition) is learned to be safe. In this way, the stimulus is treated as safe by the animal.

Forebrain: In animals with vertical posture, it is the most superior part of the brain that contains the cerebral hemispheres. In humans, it contains the cortex and subcortical structures.

Frontal: See **lobe**.

Ganglia (plural of *ganglion*): Masses of neuronal cells. Whereas the cortex is a layered structure, almost all other parts of the brain (such as the **amygdala**) are ganglionic.

Gray matter: See **matter**.

Gyrus: A ridge on the cerebral cortex, generally surrounded by one of more sulci (see **sulcus**). Gyri and sulci create the folded appearance of the brain in humans and some other mammals.

Hindbrain: Portion of the central nervous system in vertebrates that includes the medulla, pons, and cerebellum.

Homology: The same character in different animals, independent of its form and function, inherited from a common ancestor. It applies to animal parts or organs (like the brain) sharing the same building plan. Having the same embryonic origin within the building plan is one of the main criteria for identifying homology.

Lobe: The cortex is typically subdivided into macro territories called lobes: occipital (at the back of the head), temporal (near the temples and on the side of the head), parietal (at the sides and top toward the back of the head), and frontal (around the frontal part of the head). All locations are for humans.

Matter (gray or white): The gray matter is the part of brain's tissue consisting of neurons and other related cell types, such as glial cells. The white matter contains mostly long-range axons.

Medulla: The medulla oblongata or simply medulla is a long stemlike structure that makes up the lower part of the brainstem.

Midbrain: Portion of the brainstem that is closest to the **forebrain.** Among the regions discussed in the text that are in the midbrain are the superior colliculus and areas that produce the neurotransmitter dopamine (ventral tegmental area and substantia nigra).

Modularity: Degree of interdependence of the many parts that comprise a system of interest. A decomposable system can be said to be modular, whereas a nondecomposable system is not modular. More generally, modularity can be conceptualized as varying from low to high.

MRI: Magnetic resonance imaging is a technique that allows imaging the human brain (and other body parts) by picking up weak magnetic signals in the tissues of interest. Structural MRI is used to examine relatively static tissue properties, such imaging white matter. Functional MRI is typically based on measuring differences of oxygenation level when specific parts of the brain are "activated" during certain tasks or conditions.

Neocortex: Six-layered cortex in the dorsal pallium of mammals; also called *isocortex.*

Neurotransmitter: Chemicals essential for neuronal communication, typically released from the axon terminal on the presynaptic cell and which bind to the postsynaptic cell. Some common neurotransmitters include dopamine, serotonin, acetylcholine, norepinephrine, and GABA.

Nucleus: Some subcortical areas of the forebrain can be subdivided into several parts called nuclei. See figure 5.5 for some nuclei of the amygdala.

Occipital: See **lobe.**

Pallial amygdala: Component of the amygdala that is of pallial origin. In mammals, it includes the lateral, basal, and accessory basal nuclei.

Pallium: Dorsal division of the **telencephalon** (forebrain). It includes several subdivisions that are comparable across vertebrates. In mammals, its dorsal subdivision (dorsal pallium) is considerably expanded and gives rise to the cerebral cortex.

The pallium also produces during development the hippocampus and the pallial amygdala, so they are said to be of pallial origin.

Parietal: See **lobe.**

Prefrontal cortex: The most anterior part of frontal cortex (see figure 7.3).

Projection: Anatomical pathway comprised of bundles of axons that connect two areas.

Reductionism: Approach to understanding systems or mechanisms in terms of elementary parts or units. A quintessential example is working out how a mechanical clock works based on all its parts, including its many wheels.

Region: See **area.**

Spike: See **action potential.**

Striatum: Part of the subcortex (in the forebrain) consisting of the caudate and the putamen.

Subpallial amygdala: Component of the amygdala that is of subpallial origin. In mammals, it includes the central nucleus and parts of the bed nucleus of the stria terminalis.

Subpallium: Ventral division of the **telencephalon** (forebrain). It includes the **striatum** (caudate-putamen) and parts of the **amygdala** (such as the central amygdala).

Sulcus: A sulcus is a depression or groove in the cerebral cortex. It surrounds a gyrus, creating the folded appearance of the human brain. The larger sulci are called fissures. See also **gyrus.**

Synapse: The space between two neurons that permits the presynaptic neuron to pass a chemical signal to the postsynaptic cell.

System: See **complex system.**

Telencephalon: See **cerebrum.**

Temporal: See **lobe.**

Unconditioned stimulus: See **conditioned stimulus.**

Ventral striatum: The part of the striatum (caudate plus putamen) consisting of its most inferior part. At times, the term is used to refer to the nucleus accumbens. See also **accumbens.**

White matter: See **matter.**

Notes

Chapter 1

1. The full quote from the paper abstract was "we delineated 180 areas per hemisphere bounded by sharp changes in cortical architecture, function, connectivity, and/or topography."

2. Although neuroscientists more commonly use the term "area" to specify putatively well-delineated parts of the brain, I don't distinguish between the two in the book.

3. For his work on the structure of the nervous system, Ramon y Cajal was awarded the Nobel Prize in 1906.

4. Burdach actually described what is currently called the "basolateral amygdala." Other parts were added later by others.

5. When communicated by the media, neuroscience findings are almost exclusively phrased in highly modular terms. We've all heard headlines about the amygdala being the "fear center in the brain," the existence of a "reward center," as well as "spots" where memory, language, and so on take place. Whereas the media's tendency to oversimplify is perhaps unavoidable, neuroscientists are at fault, too.

6. For a related discussion, see Krakauer et al. (2017).

7. I am referring to the addiction study by Navqi et al. (2007); a study of faces by Kanwisher, McDermott, and Chun (1997); and an attention study described by Noudoost et al. (2010). The stimulation studies described by Noudoost et al. (2010) come closest to informing mechanisms.

8. The term "filler term" comes from Krakauer et al. (2017).

9. Although the quoted statement referred to "many regions," the point applies to most (if not all) brain regions.

10. Example inspired by Woodward (2013, 43).

11. Example borrowed from Striedter (2005), which was based on the work by Endler (1995).

12. Quote by Newton and Descartes from Mazzocchi (2008, 10). Overall paragraph is based on Mazzocchi (2008).

Chapter 2

1. For a brief biography of Cécile Vogt, see M. Favero, S. Mele, and T. Metitieri, "Profile of Cécile Mugnier Vogt," in WiNEu, European Women in Neuroscience, Untold Stories: The Women Pioneers of Neuroscience in Europe, 2017, http://wineurope.eu /vogt-2/.

2. From "Korbinian Brodmann," *Whonamedit? A Dictionary of Medical Eponyms*, January 9, 2015, http://www.whonamedit.com/doctor.cfm/1264.html.

3. For a discussion of the number of regions and other historical aspects, see Šimic and Hof (2015), Finger (1994), and Amunts and Zilles (2015).

4. Galen's use of "thalamus" probably referred to one of the ventricles (Rikhye, Wimmer, and Halassa 2018).

5. A century later, Leopoldo Caldini claimed proof of this role based on lesion experiments on dogs, lambs, and goats (Finger 1994, 216): "I am satisfied with having provided proof, by means of experiments and (clinical) observations, that large or small lesions of the [striatum] are followed by paralysis, more or less severe and more or less extensive."

6. See Bear, Connors, and Paradiso (2020), from which the first few paragraphs of this section draw. The figure of 86 billion neurons comes from Herculano-Houzel (2009).

7. Thiebaut de Schotten et al. (2015) is a good reference for this and the next paragraph.

Chapter 3

1. Although the original study appeared in the prestigious journal *Brain*, it was largely ignored for decades. The 1973 study, cited in the next paragraph, was by Pöppel, Held, and Frost (1973).

2. The classic reference here is by Cowey (2004).

3. Technically, it should be "superior colliculi" (plural). Throughout the text, I avoid the Latin plural, which sounds heavy-handed (for the same reason we don't say in the United States "musea" but instead "museums"). So, I don't refer to amygdalae (plural for amygdala), for example.

4. Snyder, Killackey, and Diamond (1969) and Diamond and Hall (1969) tested the visual behavior of tree shrews and squirrels after complete removal of the striate cortex. See also Day-Brown et al. (2010).

5. The classic reference is Dean, Redgrave, and Westby (1989), on which the subsequent paragraph was based. The idea that the superior colliculus is involved in defensive behaviors has not gained widespread attention. However, a growing body of findings is changing the treatment of this region as a simple sensorimotor interface. For evidence of its participation in defensive behaviors in primates, see DesJardin et al. (2013). For further discussion, see Pessoa, Medina, and Desfilis (2022).

6. In mammals, the hypothalamus also projects to intermediate layers of the superior colliculus; see Nieuwenhuys, ten Donkelaar, and Nicholson (1998), chap. 22, 1804.

7. Paragraph based directly on Dean, Redgrave, and Westby (1989, 146).

8. As newer methods become available, it now seems that a precise columnar arrangement in the PAG is an oversimplification. Instead, cells appear to integrate multiple inhibitory and excitatory inputs from distinct brain areas to select appropriate active or passive defensive behaviors (Tovote et al. 2016).

9. The connections are to the subpart of the substantia nigra called "compact" because of the dense packing of cell bodies there. For visual activation of the substantia nigra, see Comoli et al. (2003).

10. The "message is not in the molecule" is originally from chapter 5 of Thompson (2000).

11. This and next paragraphs draw directly from Thompson (2000, 140–141).

Chapter 4

1. The description in this paragraph is from Broca (1861). Online translation available at Classics in the History of Psychology, an internet resource developed by Christopher D. Green of York University, Toronto, http://psychclassics.yorku.ca/Broca/aphemie-e .htm.

2. The discussion in this and next two paragraphs builds heavily on Finger (1994, 38–40).

3. Text in this paragraph and next two paragraphs is based on Dunn and Kirsner (2003). Two key papers historically are Lashley (1952) and Teuber (1955). For an excellent modern discussion, see Young, Hilgetag, and Scannell (2000).

4. Discussion based on Plaut (1995). Patient PW was reported by Patterson and Marcel (1977) and patient CAV by Warrington (1981).

5. Yet nonmodular systems can also give rise to double dissociations when damaged; see Plaut (1995, 314).

6. Definitions by Shallice and Cooper (2011).

7. When regions A_1, A_2, etc. *jointly implement* a function F, the situation is conceptually quite different from the scenario being described. As developed in chapter 10,

we can think of the set of regions $\{A_1, A_2, \ldots\}$ as a *network* of regions that, in combination, generates the function F. I thank Marco Viola for raising these issues.

8. The example in this paragraph was described by Passingham, Stephan, and Kötter (2002).

9. An excellent, short treatment is provided by Mayr (2004).

Chapter 5

1. See the excellent paper by Parvizi (2009); see also Finger (1994, 271).

2. The historical discussion of Walter Cannon's work is based on Carroll (2016, 17–21). Further material on Cannon borrows from Finger (1994).

3. Interestingly, Bard worked independently and only collaborated with Cannon on unrelated work on the innervation of the thyroid gland, which only led to negative results and were not published (Bard 1973).

4. Bard (1934) uses the term "consciousness" explicitly.

5. For discussion and references, see Pessoa (2013, 230–231).

6. See McEwen (1998); O'Connor, O'Halloran, and Shanahan (2000); and Brosschot, Gerin, and Thayer (2006), which is the source of the chances of developing cardiovascular disease at the end of the paragraph.

7. Paragraph copied from Feinstein et al. (2013, supplemental material). Patient SM's age quoted in the subsequent paragraph is reported by Feinstein et al. (2011).

8. For example, the central amygdala projects to the dopaminergic ventral tegmental area and adjacent substantia nigra pars compacta; the noradrenergic locus coeruleus; the serotonergic raphé nuclei; and the cholinergic basal forebrain nuclei.

9. See Paré and Quirk (2017) for an outstanding discussion of related issues.

10. The debate is still not settled, but it is perhaps safe to say that it's more nuanced today. Instead of focusing on "automatic" versus "nonautomatic" processing, many studies try to determine the conditions during which emotional perception is favored and the extent to which it is.

11. Modulation of visual cortex is believed to be, in part, related to amygdala projections to visual cortex. But many other circuits are likely involved (Pessoa 2013).

12. Again, see the excellent paper by Paré and Quirk (2017).

13. See Pessoa (2010, 2013). The study of saving versus spending behavior in rhesus monkeys, discussed in the subsequent paragraph, is by Grabenhorst, Hernádi, and Schultz (2012); the human functional MRI study is by Zangemeister, Grabenhorst, and Schultz (2016).

14. As recounted by Thompson (1999, 5–6, 13).

15. Based on Olds and Milner (1954) and Olds (1958).

16. Midbrain cells that project to these diverse cortical regions are intermingled with each other (a cell that projects to the frontal cortex may be adjacent to one that projects to the temporal cortex). In addition, individual neurons often ramify, sending collateral axons to different cortical regions (say, a cell may project to both the prefrontal and parietal cortex). Overall, the projection system is rather diffuse.

17. This paragraph is close to verbatim from Schultz (2016, 24). Subsequent paragraph based on the same source.

Chapter 6

1. Macmillan (2002) covers the case of Gage extensively and argues strongly that most of what has been written about Gage is largely folklore, including changes in Gage's personality.

2. Surprisingly, the Wikipedia entry on Gage is rather carefully described and referenced; see "Phineas Gage," *Wikipedia*, M8n, citing Macmillan (2004), https://en .wikipedia.org/wiki/Phineas_Gage#M8.

3. The famous paper by James (1884) proposed what an emotion is; the quote on being "brought to consciousness" is from Lange ([1885] 1922, 75).

4. Quote by Vogt (2009, 12) is based on the original work by the neurosurgeons J. Bancaud and J. Talairach (1992).

5. See Finger (1994, 40). For example, in 1931, Otfried Foerster, one of the leaders in this area, said: "Strong faradic [electrical] stimulation produces a convulsion. . . ." (Foerster 1931, 310).

6. After years of searching for, but not finding, an English translation of Broca's monumental paper originally published in French, I sought to have it translated. With the support of the *Journal of Comparative Neurology* (a publication established in 1891, soon after Broca's work originally appeared), the paper was published in full and finally made available to English-speaking readers. See Broca (2015).

7. See Neafsey (1990, 155–156). This paper provides a good historical perspective of research on the autonomic nervous system.

8. See Vogt and Vogt (1926). The edited volume on the cingulate cortex by Vogt (2009) is an excellent source on this part of the brain.

9. See, for example, Etkin, Egner, and Kalisch (2011). The specific study is by Raczka et al. (2010).

10. See Wager et al. (2013). Since this influential initial report, the research group has published extensively on this topic.

11. A long list of neuroscientists, including myself, have made this point throughout the past decades.

12. A good source is Saper (2002), from which I draw.

13. See Craig (2002, 2009); see also Damasio (1999).

14. Descending connections are mostly found in the posterior insula (Yasui et al., 1991); however, they likely exist in the anterior insula, too.

15. Description from Koenigs and Tranel (2006).

16. "Acquired sociopathy" is discussed by Eslinger and Damasio (1985) and nicely summarized in the entry "Psychopathy," *Wikipedia*, last modified January 4, 2022, 20:19 UTC, https://en.wikipedia.org/wiki/Psychopathy.

Chapter 7

1. See Finger (1994) for historical discussion.

2. For the quote, see the translation of Broca's 1878 paper in the *Journal of Comparative Neurology* (Broca 2015, 2553). See also the commentary by Pessoa and Hof (2015).

3. See Quiroga et al. (2005); see also A. Gosline, "Why Your Brain Has a 'Jennifer Aniston Cell,'" *New Scientist*, June 22, 2005, https://www.newscientist.com/article/dn7567-why-your-brain-has-a-jennifer-aniston-cell/.

4. See the work by Moran and Desimone (1985), which was further explored in detail in the next two decades.

5. The discussion in this paragraph is from Gazzaniga, Ivry, and Mangun (2009, 426).

6. The unusual study by the French neurologist in this paragraph is by Lhermitte (1983). The quote about a "healthy jab in the buttocks" is from Gazzaniga, Ivry, and Mangun (2009, 426).

7. Discussion of the Stroop task and the Wisconsin card sort task is based on Miller and Cohen (2001). Brenda Milner's landmark study was reported in Milner (1963).

8. This paragraph is closely based on Miller and Cohen (2001, 168), including the expression "rules of the game."

9. This is one of the central ideas of the work of Stephen Grossberg; see, for example, Grossberg (2021).

Chapter 8

1. The term "intuition pumps" comes from Dennett (2013).

2. This ecosystem example is borrowed from Carroll (2017, 126).

3. This bacteria example is also borrowed from Carroll (2017, chap. 3).

4. See "Lotka-Volterra Equations," *Wikipedia*, accessed July 28, 2020, https://en .wikipedia.org/wiki/Lotka-Volterra_equations.

5. This paragraph is largely based on Juarrero (1999, 7). The term "emergence" appears to have first been proposed in the 1870s by George Henry Lewes in his book *Problems of Life and Mind* and taken up by Wilhelm Wundt in his book *Introduction to Psychology*.

6. This expression is from Deacon (2011, 166).

7. Paragraph draws from Levine et al. (2017).

8. See Bairey, Kelsic, and Kishony (2016). Just a few years ago, Levine et al. (2017, 61) pointed out that "higher-order interactions need to be demystified to become a regular part of how ecologists envision coexistence, and identifying their mechanistic basis is one way of doing so."

9. The problem of stability was central to celestial mechanics. For example, what types of trajectories do two bodies, such as the earth and the sun, exhibit? The so-called two-body problem was completely solved by Johann Bernoulli in 1734 (his brother Jacob is famous for his contributions in the field of probability, including the first version of the law of large numbers). For more than two bodies (for example, the moon, the earth, and the sun), the problem has vexed mathematicians for centuries. Remarkably, the motion of three bodies is generally nonrepeating, except in special cases. See "Three-Body Problem," *Wikipedia*, last modified January 3, 2022, 09:31 UTC, https://en.wikipedia.org/wiki/Three-body_problem, and J. Cartwright, "Physicists Discover a Whopping 13 New Solutions to Three-Body Problem," *Science*, March 8, 2013, http://www.sciencemag.org/news/2013/03/physicists-discover -whopping-13-new-solutions-three-body-problem.

10. More technically, until the 1960s, attractors were thought in terms of simple geometric subsets of the phase space (linked to the possible states of a system), like points, lines, surfaces, and simple regions of three-dimensional space (see "Attractor," *Wikipedia*, last modified October 24, 2021, 13:11 UTC, https://en.wikipedia.org /wiki/Attractor).

11. Von Bertalanffy (1950) stated that concepts like "system" and "wholeness," to which we could add "emergence" and "complexity," are vague and even somewhat mystical, and indeed many scientists displayed mistrust when faced with these concepts.

12. See also the previously mentioned "Three-Body Problem," *Wikipedia*, and J. Cartwright, "Physicists Discover a Whopping 13 New Solutions to Three-Body Problem," as noted in the discussion of Newton's interest in planetary motion.

Chapter 9

1. An exception is in the case of some reptiles, which also have a simple form of cortex with three layers.

2. "No rat was ever an ancestor of any monkey" is quoted from Hodos and Campbell (1969, 345), from which the first two paragraphs of this section are based.

3. Sources for this paragraph include Bernardi (2012), Burghardt (2013), Mikhalevich, Powell, and Logan (2017).

4. Some of the key work is by J.-P. Ewert and summarized in Ewert (1987).

5. For further discussion and primary references, see Pessoa (2018a).

6. See Pessoa et al. (2019), where many of the themes discussed in this chapter are developed in considerable depth.

7. These parts include the anterior cingulate cortex and the temporal lobe.

8. A good discussion is by Salamone and Correa (2012), from which the following paragraph is based.

9. Active models include those by Pezzulo and Cisek (2016) and Pezzulo, Rigoli, and Friston (2015).

10. In his excellent book on brain evolution, Schneider (2014, 581) proposes something very similar: "As the neocortex has become the dominant source of inputs in the mammals, it has brought all the functions of neocortex into the striatal circuitry."

11. There are many excellent treatments of the evolution of the amygdala by comparative neuroanatomists. A recent, comprehensive work is by Medina et al. (2017).

12. The connections between the basolateral amygdala and ventral striatum are very substantial; however, there are pathways to the dorsal striatum, too (Amaral et al., 1992). The pathways discussed link the intra-telencephalic circuits of the pallial amygdala with the pallial-subpallial circuits of the basal ganglia and other regions of the base of the forebrain (such as the extended amygdala).

13. A comprehensive treatment is provided by Wagner (2014).

14. The analyses accounted for overall differences in amygdala size across species. For example, humans and gorillas are considerably larger than all other animals and thus would be expected to have a larger amygdala simply based on overall body and brain sizes. See Barger et al. (2012, 2014). Concerning the comparison of the cortex, see Rilling and Insel (1999); concerning frontal lobe, see Semendeferi et al. (2002).

Chapter 10

1. The ideas in this chapter are developed more technically in Pessoa (2014, 2017b).

2. Another very influential paper was published soon after by Barabási and Albert (1999). These papers were followed by an enormous amount of research in the subsequent years.

3. Particularly useful here is the work by Kennedy and collaborators (e.g., Markov et al. 2013). For discussion of mouse and primate data, see Gămănuţ et al. (2018).

4. For an outstanding treatment of cortical-subcortical-midbrain systems, see Heimer et al. (2007); see also Pessoa et al. (2019).

5. These examples are from my lab's research; see Pessoa (2013).

6. See Bourdy and Barrot (2012). See also Nieuwenhuys, Voogd, and van Huijzen (2008) for the anatomy discussed in this section.

7. Concept developed in Pessoa (2017b).

Chapter 11

1. See "All Nobel Prizes," Nobel Prize Outreach AB 2022, *NobelPrize.org*, accessed January 8, 2022, https://www.nobelprize.org/prizes/lists/all-nobel-prizes/.

2. A good discussion is provided by Dunsmoor et al. (2015).

3. See Herry et al. (2006). For morphological changes in synapses, see Tovote et al. (2016).

4. See Herry et al. (2008) and Do-Monte et al. (2015).

5. This and the next paragraph based on Scoville and Milner (1957).

6. For the material from this and the next paragraphs, see discussions by Wikenheiser and Schoenbaum (2016), Eichenbaum et al. (2016), Jeffery (2018), and Barry and Maguire (2018). For the idea of the hippocampus and "memory map," see Buzsáki and Moser (2013).

7. Based on Maren, Phan, and Liberzon (2013); see also Maren (2011).

8. See Jones and Mishkin (1972). A good source for the material in this and the next paragraph is Schoenbaum et al. (2011).

9. Reviewed in Wikenheiser and Schoenbaum (2016).

10. See Fanselow and Lester (1988). The example of the gopher snake at the end of the paragraph is by Hirsch and Bolles (1980).

11. For an excellent recent perspective on escape behaviors, see Evans et al. (2019). See also Pessoa, Medina, and Desfilis (2022).

Chapter 12

1. See "Homologies," *Understanding Evolution*, University of California Museum of Paleontology website, June 2020, https://evolution.berkeley.edu/evolibrary/article /0_0_0/lines_04.

2. Sentence borrowed from Murray, Wise, and Graham (2017, 35): "A structure adopts new functions during evolution, yet its ancestry can be traced to something more fundamental." Discussion of the hippocampus until the end of the paragraph also borrows from their excellent treatment.

3. See Striedter (2005). The Wikipedia page on just-so-stories is actually pretty decent; see "Just-So Story," *Wikipedia*, last modified December 13, 2021, 18:52 UTC, https://en.wikipedia.org/wiki/Just-so_story.

4. Text here builds directly from Paré and Quirk (2017).

5. "Cul-de-sac" expression inspired by Kim and Jung (2018).

6. See Allen et al. (2019).

7. See "MOND Theory," *Astronoo.com*, accessed January 5, 2022, http://www.astronoo .com/en/articles/mond-theory.html.

8. See Mordehai Milgrom, "The MOND Paradigm of Modified Dynamics," *Scholarpedia* 9, no. 6 (2014): 31410, http://www.scholarpedia.org/article/The_MOND _paradigm_of_modified_dynamics.

9. The zebrafish were studied during the larval stage; see Ahrens et al. (2012).

10. See Mannino and Bressler (2015). For "mutual causality," see Hausman (1984) and Frankel (1986).

11. Computational investigations in the past years have revealed a large number of families of periodic orbits; see Šuvakov and Dmitrašinović (2003). See also, as discussed in chapter 8, "Three-Body Problem," *Wikipedia*, last modified January 3, 2022, 09:31 UTC, https://en.wikipedia.org/wiki/Three-body_problem, and J. Cartwright, "Physicists Discover a Whopping 13 New Solutions to Three-Body Problem," *Science*, March 8, 2013.

12. "Process Philosophy," *Wikipedia*, accessed August 1, 2020, https://en.wikipedia .org/wiki/Process_philosophy. Interpretation presented here is using modern terms by Rescher (2000, 3).

13. The proximity of trajectories depends on the dimensionality of the system in question (which is usually unknown) and the dimensionality of the space where data are being considered (say, after dimensionality reduction). Naturally, points projected onto a lower-dimensional representation might be closer than in the

original higher-dimensional space. For an excellent discussion of the concepts in this section, see Buonomano and Maass (2009).

14. We can arbitrarily consider "modern neuroscience" to start with Broca's clinical report (Broca 1861).

15. A good example here is quantum physics, which has immensely benefited from an intense (if at times strained) exchange between experimental physics, theoretical physics, and philosophy, for example.

References

Adolphs, R., and D. J. Anderson. 2018. *The Neuroscience of Emotion: A New Synthesis*. Princeton University Press.

Ahrens, M. B., J. M. Li, M. B. Orger, D. N. Robson, A. F. Schier, F. Engert, and R. Portugues. 2012. "Brain-wide Neuronal Dynamics during Motor Adaptation in Zebrafish." *Nature* 485 (7399): 471.

Allen, W. E., M. Z. Chen, N. Pichamoorthy, R. H. Tien, M. Pachitariu, L. Luo, and K. Deisseroth. 2019. "Thirst Regulates Motivated Behavior through Modulation of Brainwide Neural Population Dynamics." *Science* 364 (6437): 253–253.

Amaral, D. G., J. L. Price, A. Pitkanen, and S. T. Carmichael. 1992. "Anatomical Organization of the Primate Amygdaloid Complex." In *The Amygdala: Neurobiological Aspects of Emotion, Memory, and Mental Dysfunction*, edited by J. Aggleton, 1–66. Wiley-Liss.

Amir, A., P. Kyriazi, S. C. Lee, D. B. Headley, and D. Paré. 2019. "Basolateral Amygdala Neurons Are Activated during Threat Expectation." *Journal of Neurophysiology* 121 (5): 1761–1777.

Amunts, K., and K. Zilles. 2015. "Architectonic Mapping of the Human Brain beyond Brodmann." *Neuron* 88 (6): 1086–1107.

Anderson, M. L., J. Kinnison, and L. Pessoa. 2013. "Describing Functional Diversity of Brain Regions and Brain Networks." *Neuroimage* 73:50–58.

Averbeck, B. B., and M. Seo. 2008. "The Statistical Neuroanatomy of Frontal Networks in the Macaque." *PLoS Computational Biology* 4 (4): e1000050.

Bairey, E., E. D. Kelsic, and R. Kishony. 2016. "Higher-Order Species Interactions Shape Ecosystem Diversity." *Nature Communications* 7 (1): 1–7.

Bancaud J., and J. Talairach 1992. "Clinical Semiology of Frontal Lobe Seizures." *Advances in Neurology* 57:3–58.

Bandler, R., and M. T. Shipley. 1994. "Columnar Organization in the Midbrain Periaqueductal Gray: Modules for Emotional Expression?" *Trends in Neurosciences* 17 (9): 379–389.

Barabási, A. L., and R. Albert. 1999. "Emergence of Scaling in Random Networks." *Science* 286 (5439): 509–512.

Bard, P. 1934. "On Emotional Expression after Decortication with Some Remarks on Certain Theoretical Views: Part I." *Psychological Review* 41 (4): 309.

Bard, P. 1973. "The Ontogenesis of One Physiologist." *Annual Review of Physiology* 35 (1): 1–16.

Barger, N., K. L. Hanson, K. Teffer, N. M. Schenker-Ahmed, and K. Semendeferi. 2014. "Evidence for Evolutionary Specialization in Human Limbic Structures." *Frontiers in Human Neuroscience* 8:277.

Barger, N., L. Stefanacci, C. M. Schumann, C. C. Sherwood, J. Annese, J. M. Allman, J. A. Buckwalter, P. R. Hof, and K. Semendeferi. 2012. "Neuronal Populations in the Basolateral Nuclei of the Amygdala Are Differentially Increased in Humans Compared with Apes: A Stereological Study." *Journal of Comparative Neurology* 520 (13): 3035–3054.

Barry, D. N., and E. A. Maguire. 2018. "Remote Memory and the Hippocampus: A Constructive Critique." *Trends in Cognitive Sciences* 23 (2): 128–142.

Bear, M. F., B. W. Connors, and M. A. Paradiso. 2020. *Neuroscience: Exploring the Brain*. 4th ed. Jones & Bartlett Learning.

Bechtel, W. 2008. *Mental Mechanisms: Philosophical Perspectives on Cognitive Neuroscience*. Taylor & Francis.

Bechtel, W., and R. C. Richardson. 2010. *Discovering Complexity: Decomposition and Localization as Strategies in Scientific Research*. MIT Press.

Bernardi, G. 2012. "The Use of Tools by Wrasses (Labridae)." *Coral Reefs* 31 (1): 39–39.

Bourdy, R., and M. Barrot. 2012. "A New Control Center for Dopaminergic Systems: Pulling the VTA by the Tail." *Trends in Neurosciences* 35 (11): 681–690.

Brandão, M. L., V. Z. Anseloni, J. E. Pandossio, J. E. De Araujo, and V. M. Castilho. 1999. "Neurochemical Mechanisms of the Defensive Behavior in the Dorsal Midbrain." *Neuroscience & Biobehavioral Reviews* 23 (6): 863–875.

Broca, P. 1861. "Remarques sur le siège de la faculté du langage articulé, suivies d'une observation d'aphémie (perte de la parole)." *Bulletin de la Société Anatomique* 6:330–357.

Broca, P. 2015. "Comparative Anatomy of the Cerebral Convolutions: The Great Limbic Lobe and the Limbic Fissure in the Mammalian Series." Translated by D. Furlani. *Journal of Comparative Neurology* 523 (17): 2501–2554.

Brosschot, J. F., W. Gerin, and J. F. Thayer. 2006. "The Perseverative Cognition Hypothesis: A Review of Worry, Prolonged Stress-Related Physiological Activation, and Health." *Journal of Psychosomatic Research* 60 (2): 113–124.

Brosschot, J. F., B. Verkuil, and J. F. Thayer. 2018. "Generalized Unsafety Theory of Stress: Unsafe Environments and Conditions, and the Default Stress Response." *International Journal of Environmental Research and Public Health* 15 (3): 464.

Buonomano, D. V., and W. Maass. 2009. "State-Dependent Computations: Spatiotemporal Processing in Cortical Networks." *Nature Reviews Neuroscience* 10 (2): 113–125.

Burghardt, G. M. 2013. "Environmental Enrichment and Cognitive Complexity in Reptiles and Amphibians: Concepts, Review, and Implications for Captive Populations." *Applied Animal Behaviour Science* 147 (3–4): 286–298.

Buzsáki, G., and E. I. Moser. 2013. "Memory, Navigation and Theta Rhythm in the Hippocampal-Entorhinal System." *Nature Neuroscience* 16:130–138.

Cannon, W. B., and S. W. Britton. 1925. "Studies on the Conditions of Activity in Endocrine Glands: XV. Pseudaffective Medulliadrenal Secretion." *American Journal of Physiology* 72 (2): 283–294.

Card, J. P., L. W. Swanson, and R. Y. Moore. 2003. "The Hypothalamus: An Overview of Regulatory Systems." In *Fundamental Neuroscience*, 2nd ed., edited by L. R. Squire, F. E. Bloom, S. K. McConnell, J. L. Roberts, N. C. Spitzer, and M. J. Zigmond, 897–909. Academic Press.

Carroll, S. B. 2017. *The Serengeti Rules: The Quest to Discover How Life Works and Why It Matters*. Princeton University Press.

Chareyron, L. J., P. Banta Lavenex, D. G. Amaral, and P. Lavenex. 2011. "Stereological Analysis of the Rat and Monkey Amygdala." *Journal of Comparative Neurology* 519 (16): 3218–3239.

Comoli, E., V. Coizet, J. Boyes, J. P. Bolam, N. S. Canteras, R. H. Quirk, P. G. Overton, and P. Redgrave. 2003. "A Direct Projection from Superior Colliculus to Substantia Nigra for Detecting Salient Visual Events." *Nature Neuroscience* 6 (9): 974.

Constantinidis, C., M. N. Franowicz, and P. S. Goldman-Rakic. 2001. "The Sensory Nature of Mnemonic Representation in the Primate Prefrontal Cortex." *Nature Neuroscience* 4 (3): 311–316.

Cowey, A. 2004. "The 30th Sir Frederick Bartlett Lecture: Fact, Artefact, and Myth about Blindsight." *Quarterly Journal of Experimental Psychology Section A* 57 (4): 577–609.

Craig, A. D. 2002. "How Do You Feel? Interoception: The Sense of the Physiological Condition of the Body. *Nature Reviews Neuroscience* 3 (8): 655–666.

Craig, A. D. 2009. "How Do You Feel—Now? The Anterior Insula and Human Awareness." *Nature Reviews Neuroscience* 10 (1): 59–70.

Crow, T. J. 1980. "Molecular Pathology of Schizophrenia: More than One Disease Process?" *British Medical Journal* 280 (6207): 66–68.

Damasio, A. 1979. "The Frontal Lobes." In *Clinical Neuropsychology*, edited by K. M. Heilman and E. Valenstein, 360–412. Oxford University Press.

Damasio, A. R. 1999. *The Feeling of What Happens: Body and Emotion in the Making of Consciousness*. Houghton Mifflin Harcourt.

Davis, M. 1992. "The Role of the Amygdala in Fear and Anxiety." *Annual Review of Neuroscience* 15 (1): 353–375.

Day-Brown, J. D., H. Wei, R. D. Chomsung, H. M. Petry, and M. E. Bickford. 2010. "Pulvinar Projections to the Striatum and Amygdala in the Tree Shrew." *Frontiers in Neuroanatomy* 4:143.

Deacon, T. W. 2011. *Incomplete Nature: How Mind Emerged from Matter*. W. W. Norton.

Dean, P., P. Redgrave, and G. M. Westby. 1989. "Event or Emergency? Two Response Systems in the Mammalian Superior Colliculus." *Trends in Neurosciences* 12 (4): 137–147.

Dennett, D. C. 2013. *Intuition Pumps and Other Tools for Thinking*. W. W. Norton.

DesJardin, J. T., A. L. Holmes, P. A. Forcelli, C. E. Cole, J. T. Gale, L. L. Wellman, K. Gale, and L. Malkova. 2013. "Defense-like Behaviors Evoked by Pharmacological Disinhibition of the Superior Colliculus in the Primate." *Journal of Neuroscience* 33 (1): 150–155.

Diamond, I. T., and W. C. Hall. 1969. "Evolution of Neocortex." *Science* 164:251–262.

Do-Monte, F. H., G. Manzano-Nieves, K. Quiñones-Laracuente, L. Ramos-Medina, and G. J. Quirk. 2015. "Revisiting the Role of Infralimbic Cortex in Fear Extinction with Optogenetics." *Journal of Neuroscience* 35 (8): 3607–3615.

Dugas-Ford, J., and C. W. Ragsdale. 2015. "Levels of Homology and the Problem of Neocortex." *Annual Review of Neuroscience* 38:351–368.

Dunbar, K., and D. Sussman. 1995. "Toward a Cognitive Account of Frontal Lobe Function: Simulating Frontal Lobe Deficits in Normal Subjects." *Annals of the New York Academy of Sciences* 769:289–304.

Dunn, J. C., and K. Kirsner. 2003. "What Can We Infer from Double Dissociations?" *Cortex* 39 (1): 1–8.

Dunsmoor, J. E., Y. Niv, N. Daw, and E. A. Phelps. 2015. "Rethinking Extinction." *Neuron* 88 (1): 47–63.

Dupré, J. A., and D. J. Nicholson. 2018. "A Manifesto for a Processual Philosophy of Biology." In *Everything Flows: Towards a Processual Philosophy of Biology*, edited by D. J. Nicholson and J. Dupré, 3–45. Oxford University Press.

Edinger, L. 1908. "The Relations of Comparative Anatomy to Comparative Psychology." Translated by H. W. Rand. *Journal of Comparative Neurology and Psychology* 53 (5): 437–457.

Eichenbaum, H., D. G. Amaral, E. A. Buffalo, G. Buzsáki, N. Cohen, L. Davachi, L. Frank, S. Heckers, R.G.M Morris, E. I Moser, L. Nadel, J. O'Keefe, A. Preston, C. Ranganath, A. Silva, and M. Witter. 2016. "Hippocampus at 25." *Hippocampus* 26 (10): 1238–1249.

Ellis, A. W. 1987. "Intimations of Modularity, Or, the Modelarity of Mind: Doing Cognitive Neuropsychology without Syndromes." In *The Cognitive Neuropsychology of Language*, edited by M. Coltheart, G. Sartori, and R. Job, 397–408. Lawrence Erlbaum Associates.

Emery, N. J., and N. S. Clayton. 2004. "The Mentality of Crows: Convergent Evolution of Intelligence in Corvids and Apes." *Science* 306 (5703): 1903–1907.

Endler, J. A. 1995. "Multiple-Trait Coevolution and Environmental Gradients in Guppies." *Trends in Ecology & Evolution* 10 (1): 22–29.

Eslinger, P. J., and A. R. Damasio. 1985. "Severe Disturbance of Higher Cognition after Bilateral Frontal Lobe Ablation: Patient EVR." *Neurology* 35 (12): 1731–1741.

Etkin, A., T. Egner, and R. Kalisch. 2011. "Emotional Processing in Anterior Cingulate and Medial Prefrontal Cortex." *Trends in Cognitive Sciences* 15 (2): 85–93.

Evans, D. A., A. V. Stempel, R. Vale, and T. Branco. 2019. "Cognitive Control of Escape Behaviour." *Trends in Cognitive Sciences* 23 (4): 334–348.

Ewert, J.-P. 1987. "Neuroethology: Toward a Functional Analysis of Stimulus-Response Mediating and Modulating Neural Circuitries." In *Cognitive Processes and Spatial Orientation in Animal and Man*, vol. 1, *Experimental Animal Psychology and Ethology*, edited by P. Ellen and C. Thinus-Blanc, 177–200. Springer.

Fanselow, M. S., and L. S. Lester. 1988. "A Functional Behavioristic Approach to Aversively Motivated Behavior: Predatory Imminence as a Determinant of the Topography of Defensive Behavior." In *Evolution and Learning*, edited by R. C. Bolles and M. D. Beecher, 185–212. Lawrence Elbaum Associates.

Feinstein, J. S., R. Adolphs, A. Damasio, and D. Tranel. 2011. "The Human Amygdala and the Induction and Experience of Fear." *Current Biology* 21 (1): 34–38.

Feinstein, J. S., C. Buzza, R. Hurlemann, R. L. Follmer, N. S. Dahdaleh, W. H. Coryell, M. J. Welsh, D. Tranel, and J. A. Wemmie. 2013. "Fear and Panic in Humans with Bilateral Amygdala Damage." *Nature Neuroscience* 16 (3): 270–272.

Finger, S. 1994. *Origins of Neuroscience: A History of Explorations into Brain Function.* Oxford University Press.

Foerster, O. 1931. "The Cerebral Cortex in Man." *Lancet* 2:309–312.

Frankel, L. 1986. "Mutual Causation, Simultaneity and Event Description." *Philosophical Studies* 49 (3): 361–372.

Gămănuţ, R., H. Kennedy, Z. Toroczkai, M. Ercsey-Ravasz, D. C. Van Essen, K. Knoblauch, and A. Burkhalter. 2018. "The Mouse Cortical Connectome, Characterized by an Ultra-Dense Cortical Graph, Maintains Specificity by Distinct Connectivity Profiles." *Neuron* 97 (3): 698–715.

Gazzaniga, M. S., R. B. Ivry, and G. R. Mangun. 2009. *Cognitive Neuroscience: The Biology of the Mind*. W. W. Norton.

Genon, S., A. Reid, R. Langner, K. Amunts, and S. B. Eickhoff. 2018. "How to Characterize the Function of a Brain Region." *Trends in Cognitive Sciences* 22 (4): 350–364.

Glasser, M. F., T. S. Coalson, E. C. Robinson, C. D. Hacker, J. Harwell, E. Yacoub, K. Ugurbil, J. Andersson, C. F. Beckmann, M. Jenkinson, S. M. Smith, and D. C. Van Essen. 2016. "A Multi-Modal Parcellation of Human Cerebral Cortex." *Nature* 536 (7615): 171–178.

Goldman-Rakic, P. S. 1988. "Topography of Cognition: Parallel Distributed Networks in Primate Association Cortex." *Annual Review of Neuroscience* 11 (1): 137–156.

Grabenhorst, F., I. Hernádi, and W. Schultz. 2012. "Prediction of Economic Choice by Primate Amygdala Neurons." *Proceedings of the National Academy of Sciences* 109 (46): 18950–18955.

Grayson, D. S., E. Bliss-Moreau, C. J. Machado, J. Bennett, K. Shen, K. A. Grant, D.A. Fair, and D. G. Amaral. 2016. "The Rhesus Monkey Connectome Predicts Disrupted Functional Networks Resulting from Pharmacogenetic Inactivation of the Amygdala." *Neuron* 91 (2): 453–466.

Grossberg, S. 2021. *Conscious Mind, Resonant Brain: How Each Brain Makes a Mind*. Oxford University Press.

Gründemann, J., Y. Bitterman, T. Lu, S. Krabbe, B. F. Grewe, M. J. Schnitzer, and A. Lüthi. 2019. "Amygdala Ensembles Encode Behavioral States." *Science* 364:6437.

Guimera, R., and L. Nunes Amaral. (2005). "Functional Cartography of Complex Metabolic Networks." *Nature* 433 (7028): 895–900.

Hausman, D. M. 1984. "Causal Priority." *Noûs* 18 (2): 261–279.

Hebb, D. O. 1949. *The Organization of Behavior: A Neuropsychological Theory*. John Wiley and Chapman & Hall.

Heimer, L., G. W. Van Hoesen, M. Trimble, and D. S. Zahm. 2007. *Anatomy of Neuropsychiatry: The New Anatomy of the Basal Forebrain and Its Implications for Neuropsychiatric Illness*. Academic Press.

Herculano-Houzel, S. 2009. "The Human Brain in Numbers: A Linearly Scaled-Up Primate Brain." *Frontiers in Human Neuroscience* 3 (November).

Herry, C., S. Ciocchi, V. Senn, L. Demmou, C. Müller, and A. Lüthi. 2008. "Switching On and Off Fear by Distinct Neuronal Circuits." *Nature* 454 (7204): 600–606.

Herry, C., P. Trifilieff, J. Micheau, A. Luthi, and N. Mons. 2006. "Extinction of Auditory Fear Conditioning Requires MAPK/ERK Activation in the Basolateral Amygdala." *European Journal of Neuroscience* 24:261–269.

Hirsch, S. M., and R. C. Bolles. 1980. "On the Ability of Prey to Recognize Predators." *Zeitschrift für Tierpsychologie* 54 (1): 71–84.

Hodos, W., and C.B.G. Campbell. 1969. "Scala Naturae: Why There Is No Theory in Comparative Psychology." *Psychological Review* 76 (4): 337–350.

Holstege, G. 1991. "Descending Pathways from the Periaqueductal Gray and Adjacent Areas." In *The Midbrain Periaqueductal Gray Matter*, edited by A. Depaulis and R. Bandler, 239–265. Plenum Press.

James, W. 1884. "What Is an Emotion?" *Mind* 9 (34): 188–205.

Jeffery, K. J. (2018). "The Hippocampus: From Memory, to Map, to Memory Map." *Trends in Neurosciences* 41 (2): 64–66.

Jones, B., and M. Mishkin. 1972. "Limbic Lesions and the Problem of Stimulus—Reinforcement Associations." *Experimental Neurology* 36 (2): 362–377.

Juarrero, A. 1999. *Dynamics in Action: Intentional Behavior as a Complex System.* MIT Press.

Kaada, B. R. 1960. "Cingulate, Posterior Orbital, Anterior Insular and Temporal Pole Cortex." In *Handbook of Physiology: Neurophysiology II*, edited by J. Field and H. W. Magoun, 1345–1371. American Physiological Society.

Kaada, B. R. 1972. "Stimulation and Regional Ablation of the Amygdaloid Complex with Reference to Functional Representations." In *The Neurobiology of the Amygdala*, edited by B. E. Eleftheriou, 205–281. Plenum Press.

Kanwisher, N., J. McDermott, and M. M. Chun. 1997. "The Fusiform Face Area: A Module in Human Extrastriate Cortex Specialized for Face Perception." *Journal of Neuroscience* 17 (11): 4302–4311.

Kapp, B. S., R. C. Frysinger, M. Gallagher, and J. R. Haselton. 1979. "Amygdala Central Nucleus Lesions: Effect on Heart Rate Conditioning in the Rabbit. *Physiology & Behavior* 23 (6): 1109–1117.

Kim, J. J., and M. W. Jung. 2018. "Fear Paradigms: The Times They Are a-Changin'." *Current Opinion in Behavioral Sciences* 24:38–43.

Koenigs, M., and D. Tranel. 2006. "Pseudopsychopathy: A Perspective from Cognitive Neuroscience." In *The Orbitofrontal Cortex*, edited by D. Zald and S. Rauch, 597–619. Oxford University Press.

Krakauer, J. W., A. A. Ghazanfar, A. Gomez-Marin, M. A. MacIver, and D. Poeppel. 2017. "Neuroscience Needs Behavior: Correcting a Reductionist Bias." *Neuron* 93 (3): 480–490.

Lange, C. G. (1885) 1922. "The Emotions: A Psychophysiological Study." In *Psychology Classics: A Series of Reprints and Translations*, edited by K. Dunlap, translated by I. A. Haupt, 33–90. Williams & Wilkins.

Lashley, K. S. 1952. "Functional Interpretation of Anatomic Patterns." *Research Publications Association for Research in Nervous and Mental Disease* 30:529–547.

Levine, J. M., J. Bascompte, P. B. Adler, and S. Allesina. 2017. "Beyond Pairwise Mechanisms of Species Coexistence in Complex Communities." *Nature* 546 (7656): 56–64.

Lhermitte, F. 1983. "'Utilization Behaviour' and Its Relation to Lesions of the Frontal Lobes." *Brain* 106 (2): 237–255.

Lim, S. L., S. Padmala, and L. Pessoa. 2009. "Segregating the Significant from the Mundane on a Moment-to-Moment Basis via Direct and Indirect Amygdala Contributions." *Proceedings of the National Academy of Sciences* 106 (39): 16841–16846.

Macmillan, M. 2002. *An Odd Kind of Fame: Stories of Phineas Gage*. MIT Press.

Macmillan, M. B. 2004. "Inhibition and Phineas Gage: Repression and Sigmund Freud." *Neuropsychoanalysis* 6 (2): 181–192.

Mannino, M., and S. L. Bressler. 2015. "Foundational Perspectives on Causality in Large-Scale Brain Networks." *Physics of Life Reviews* 15:107–123.

Maren, S. 2011. "Seeking a Spotless Mind: Extinction, Deconsolidation, and Erasure of Fear Memory." *Neuron* 70 (5): 830–845.

Maren, S., K. L. Phan, and I. Liberzon. 2013. "The Contextual Brain: Implications for Fear Conditioning, Extinction, and Psychopathology." *Nature Reviews Neuroscience* 14 (6): 417–428.

Markov, N. T., M. Ercsey-Ravasz, D. C. Van Essen, K. Knoblauch, Z. Toroczkai, and H. Kennedy. 2013. "Cortical High-Density Counterstream Architectures." *Science* 342 (6158).

Mayr, E. 2004. *What Makes Biology Unique? Considerations on the Autonomy of a Scientific Discipline*. Cambridge University Press.

Mazzocchi, F. 2008. "Complexity in Biology." *EMBO Reports* 9 (1): 10–14.

McEwen, B. S. 1998. "Stress, Adaptation, and Disease: Allostasis and Allostatic Load." *Annals of the New York Academy of Sciences* 840 (1): 33–44.

Medina, L., A. Abellán, A. Vicario, B. Castro-Robles, and E. Desfilis. 2017. "The Amygdala." In *Evolution of Nervous Systems*. Vol. 2, edited by J. Kaas, 427–478. Elsevier.

Mesulam, M. M. 1981. "A Cortical Network for Directed Attention and Unilateral Neglect." *Annals of Neurology* 10 (4): 309–325.

Mikhalevich, I., R. Powell, and C. Logan. 2017. "Is Behavioural Flexibility Evidence of Cognitive Complexity? How Evolution Can Inform Comparative Cognition." *Interface Focus* 7 (3): 20160121.

Miller, E. K., and J. D. Cohen. 2001. "An Integrative Theory of Prefrontal Cortex Function." *Annual Review of Neuroscience* 24 (1): 167–202.

Milner, B. 1963. "Effects of Different Brain Lesions on Card Sorting: The Role of the Frontal Lobes." *Archives of Neurology* 9 (1): 90–100.

Modha, D. S., and R. Singh. 2010. "Network Architecture of the Long-Distance Pathways in the Macaque Brain." *Proceedings of the National Academy of Sciences* 107 (30): 13485–13490.

Moran, J., and R. Desimone. 1985. "Selective Attention Gates Visual Processing in the Extrastriate Cortex." *Science* 229 (4715): 782–784.

Morgan, M. A., L. M. Romanski, and J. E. LeDoux. 1993. "Extinction of Emotional Learning: Contribution of Medial Prefrontal Cortex." *Neuroscience Letters* 163 (1): 109–113.

Morgane, P. J. 1979. "Historical and Modern Concepts of Hypothalamic Organization and Function." In *Handbook of the Hypothalamus: Vol. 1: Anatomy of the Hypothalamus*, edited by P. J. Morgane and J. Panksepp, 1–64. Marcel Dekker.

Murray, E. A., S. P. Wise, and K. S. Graham. 2017. *The Evolution of Memory Systems: Ancestors, Anatomy, and Adaptations.* Oxford University Press.

Najafi, M., B. W. McMenamin, J. Z. Simon, and L. Pessoa. 2016. "Overlapping Communities Reveal Rich Structure in Large-Scale Brain Networks during Rest and Task Conditions." *Neuroimage* 135:92–106.

Naqvi, N. H., D. Rudrauf, H. Damasio, and A. Bechara. 2007. "Damage to the Insula Disrupts Addiction to Cigarette Smoking." *Science* 315 (5811): 531–534.

Neafsey, E. J. 1990. "Prefrontal Cortical Control of the Autonomic Nervous System: Anatomical and Physiological Observations." *Progress in Brain Research* 85:147–166.

Nieuwenhuys, R., H. J. ten Donkelaar, and C. Nicholson. 1998. *The Central Nervous System of Vertebrates.* Vol. 3. Springer-Verlag.

Nieuwenhuys, R., J. Voogd, and C. van Huijzen. 2008. *The Human Central Nervous System: A Synopsis and Atlas.* 4th ed. Springer Science and Business Media.

Noudoost, B., M. H. Chang, N. A. Steinmetz, and T. Moore. 2010. "Top-Down Control of Visual Attention." *Current Opinion in Neurobiology* 20 (2): 183–190.

O'Connor, T. M., D. J. O'Halloran, and F. Shanahan. 2000. "The Stress Response and the Hypothalamic-Pituitary-Adrenal Axis: From Molecule to Melancholia." *QJM: An International Journal of Medicine* 93 (6): 323–333.

Olds, J. 1958. "Self-Stimulation of the Brain: Its Use to Study Local Effects of Hunger, Sex, and Drugs." *Science* 127 (3294): 315–324.

Olds, J., and P. Milner. 1954. "Positive Reinforcement Produced by Electrical Stimulation of Septal Area and Other Regions of Rat Brain." *Journal of Comparative and Physiological Psychology* 47 (6): 419.

Palla, G., I. Derényi, I. Farkas, and T. Vicsek. 2005. "Uncovering the Overlapping Community Structure of Complex Networks in Nature and Society." *Nature* 435 (7043): 814–818.

Paré, D., and G. J. Quirk. 2017. "When Scientific Paradigms Lead to Tunnel Vision: Lessons from the Study of Fear." *npj Science of Learning* 2 (1): 1–8.

Parvizi, J. 2009. "Corticocentric Myopia: Old Bias in New Cognitive Sciences." *Trends in Cognitive Sciences* 13 (8): 354–359.

Passingham, R. E., K. E. Stephan, and R. Kötter. 2002. "The Anatomical Basis of Functional Localization in the Cortex." *Nature Reviews Neuroscience* 3 (8): 606–616.

Patterson, K. E., and A. J. Marcel. 1977. "Aphasia, Dyslexia and the Phonological Coding of Written Words." *Quarterly Journal of Experimental Psychology* 29 (2): 307–318.

Pavlov, I. P. 1927. *Conditioned Reflexes: An Investigation of the Physiological Activity of the Cerebral Cortex.* Translated by G. V. Anrep. Oxford University Press.

Pessoa, L. 2010. "Emotion and Cognition and the Amygdala: From 'What Is It?' to 'What's to Be Done?'" *Neuropsychologia* 48 (12): 3416–3429.

Pessoa, L. 2013. *The Cognitive-Emotional Brain: From Interactions to Integration.* MIT Press.

Pessoa, L. 2014. "Understanding Brain Networks and Brain Organization." *Physics of Life Reviews* 11 (3): 400–435.

Pessoa, L. 2017a. "Cognitive-Motivational Interactions: Beyond Boxes-and-Arrows Models of the Mind-Brain." *Motivation Science* 3 (3): 287–303.

Pessoa, L. 2017b. "A Network Model of the Emotional Brain." *Trends in Cognitive Sciences,* 21 (5): 357–371.

Pessoa, L. 2018a. "Emotion and the Interactive Brain: Insights from Comparative Neuroanatomy and Complex Systems." *Emotion Review* 10 (3): 204–216.

Pessoa, L. 2018b. "Understanding Emotion with Brain Networks." *Current Opinion in Behavioral Sciences* 19:19–25.

Pessoa, L., and R. Adolphs. 2010. "Emotion Processing and the Amygdala: From a 'Low Road' to 'Many Roads' of Evaluating Biological Significance." *Nature Reviews Neuroscience*, 11 (11): 773–782.

Pessoa, L., E. Gutierrez, P. A. Bandettini, and L. G. Ungerleider. 2002. "Neural Correlates of Visual Working Memory: fMRI Amplitude Predicts Task Performance." *Neuron* 35 (5): 975–987.

Pessoa, L., and P. R. Hof. 2015. "From Paul Broca's Great Limbic Lobe to the Limbic System." *Journal of Comparative Neurology* 523(17): 2495–2500.

Pessoa, L., L. Medina, and E. Desfilis. 2022. "Refocusing Neuroscience: Moving Away from Mental Categories and toward Complex Behaviors." *Proceedings of the Transactions of the Royal Society B* 377 (1844): 20200534.

Pessoa, L., L. Medina, P. R. Hof, and E. Desfilis. 2019. "Neural Architecture of the Vertebrate Brain: Implications for the Interaction between Emotion and Cognition." *Neuroscience & Biobehavioral Reviews* 107:296–312.

Pezzulo, G., and P. Cisek. 2016. "Navigating the Affordance Landscape: Feedback Control as a Process Model of Behavior and Cognition." *Trends in Cognitive Sciences* 20 (6): 414–424.

Pezzulo, G., F. Rigoli, and K. Friston. 2015. "Active Inference, Homeostatic Regulation and Adaptive Behavioral Control." *Progress in Neurobiology* 134:17–35.

Plaut, D. C. 1995. "Double Dissociation without Modularity: Evidence from Connectionist Neuropsychology." *Journal of Clinical and Experimental Neuropsychology* 17 (2): 291–321.

Pöppel, E., R. Held, and D. Frost. 1973. "Residual Visual Function after Brain Wounds Involving the Central Visual Pathways in Man." *Nature* 243 (5405): 295–296.

Pribram, K. H. 1960. "A Review of Theory in Physiological Psychology." *Annual Review of Psychology* 11 (1): 1–40.

Price, C. J., and K. J. Friston. 2005. "Functional Ontologies for Cognition: The Systematic Definition of Structure and Function." *Cognitive Neuropsychology* 22 (3–4): 262–275.

Quiroga, R. Q., L. Reddy, G. Kreiman, C. Koch, and I. Fried. 2005. "Invariant Visual Representation by Single Neurons in the Human Brain." *Nature* 435 (7045): 1102–1107.

Raczka, K. A., N. Gartmann, M.-L. Mechias, A. Reif, C. Büchel, J. Deckert, and R. Kalisch. 2010. "A Neuropeptide S Receptor Variant Associated with Overinterpretation of Fear Reactions: A Potential Neurogenetic Basis for Catastrophizing." *Molecular Psychiatry* 15:1067–1074.

Rescher, N. (2000). *Process Philosophy: A Survey of Basic Issues*. University of Pittsburgh Press.

Riddoch, G. 1917. "Dissociation of Visual Perceptions due to Occipital Injuries, with Especial Reference to Appreciation of Movement." *Brain* 40 (1): 15–57.

Rikhye, R. V., R. D. Wimmer, and M. M. Halassa. 2018. "Toward an Integrative Theory of Thalamic Function." *Annual Review of Neuroscience* 41:163–183.

Rilling, J. K., and T. R. Insel. 1999. "The Primate Neocortex in Comparative Perspective Using Magnetic Resonance Imaging." *Journal of Human Evolution* 37 (2): 191–223.

Roesch, M., and G. Schoenbaum. 2006. "From Associations to Expectancies: Orbitofrontal Cortex as Gateway between the Limbic System and Representational Memory." In *The Orbitofrontal Cortex*, edited by D. H. Zald and S. L. Rauch, 199–236. Oxford University Press.

Salamone, J. D., and M. Correa. 2012. "The Mysterious Motivational Functions of Mesolimbic Dopamine." *Neuron* 76 (3): 470–485.

Saper, C. B. 2002. "The Central Autonomic Nervous System: Conscious Visceral Perception and Autonomic Pattern Generation." *Annual Review of Neuroscience* 25 (1): 433–469.

Schneider, G. E. 2014. *Brain Structure and Its Origins: In Development and in Evolution of Behavior and the Mind*. MIT Press.

Schoenbaum, G., Y. Takahashi, T. L. Liu, and M. A. McDannald. 2011. "Does the Orbitofrontal Cortex Signal Value?" *Annals of the New York Academy of Sciences* 1239:87–99.

Schultz, W. 2016. "Dopamine Reward Prediction Error Coding." *Dialogues in Clinical Neuroscience* 18 (1): 23–32.

Scoville, W. B., and B. Milner. 1957. "Loss of Recent Memory after Bilateral Hippocampal Lesions." *Journal of Neurology, Neurosurgery, and Psychiatry* 20 (1): 11–21.

Semendeferi, K., A. Lu, N. Schenker, and H. Damasio. 2002. "Humans and Great Apes Share a Large Frontal Cortex." *Nature Neuroscience* 5 (3): 272–276.

Shallice, T., and R. Cooper. 2011. *The Organization of Mind*. Oxford University Press.

Shepherd, G. M. 2011. "The Microcircuit Concept Applied to Cortical Evolution: From Three-Layer to Six-Layer Cortex." *Frontiers in Neuroanatomy* 5 (30): 1–15.

Shipp, S. 2017. "The Functional Logic of Corticostriatal Connections." *Brain Structure and Function* 222 (2): 669–706.

Silva, B. A., C. Mattucci, P. Krzywkowski, E. Murana, A. Illarionova, V. Grinevich, N. S. Canteras, D. Ragozzino, and C. T. Gross. 2013. "Independent Hypothalamic Circuits for Social and Predator Fear." *Nature Neuroscience* 16 (12): 1731–1733.

Šimic, G., and P. R. Hof. 2015. "In Search of the Definitive Brodmann's Map of Cortical Areas in Human." *Journal of Comparative Neurology* 523 (1): 5–14.

Snyder, M., H. Killackey, and I. T. Diamond. 1969. "Color Vision in the Tree Shrew after Removal of Posterior Neocortex." *Journal of Neurophysiology* 32:554–563.

Strehler, B. L. 1991. "Where Is the Self? A Neuroanatomical Theory of Consciousness." *Synapse* 7 (1): 44–91.

Striedter, G. F. 2005. *Principles of Brain Evolution*. Sinauer Associates.

Stringer, C., M. Pachitariu, N. Steinmetz, C. B. Reddy, M. Carandini, and K. D. Harris. 2019. "Spontaneous Behaviors Drive Multidimensional, Brainwide Activity." *Science* 364 (6437): 255–255.

Šuvakov, M., and V. Dmitrašinović. 2013. "Three Classes of Newtonian Three-Body Planar Periodic Orbits." *Physical Review Letters* 110 (11): 114301.

Teuber, H. L. 1955. "Physiological Psychology." *Annual Review of Psychology* 6 (1): 267–296.

Thiebaut de Schotten, M., F. Dell'Acqua, P. Ratiu, A. Leslie, H. Howells, E. Cabanis, M. T. Iba-Zizen, O. Plaisant, A. Simmons, N. F. Dronkers, S. Corkin, and M. Catani. 2015. "From Phineas Gage and Monsieur Leborgne to HM: Revisiting Disconnection Syndromes." *Cerebral Cortex* 25 (12): 4812–4827.

Thompson, R. F. 1999. "James Olds: 1922–1976, a Biographical Memoir." In *Biographical Memoirs: Volume 77*, 246–263. National Academy Press.

Thompson, R. F. 2000. *The Brain: A Neuroscience Primer*. Macmillan.

Thorpe, S. J., E. T. Rolls, and S. Maddison. 1983. "The Orbitofrontal Cortex: Neuronal Activity in the Behaving Monkey." *Experimental Brain Research* 49 (1): 93–115.

Tovote, P., M. S. Esposito, P. Botta, F. Chaudun, J. P. Fadok, M. Markovic, S.B.E. Wolff, C. Ramakrishnan, L. Fenno, K. Deisseroth, C. Herry, S. Arber, and A. Lüthi. 2016. "Midbrain Circuits for Defensive Behaviour." *Nature* 534 (7606): 206–212.

Tyszka, J. M., D. P. Kennedy, R. Adolphs, and L. K. Paul. 2011. "Intact Bilateral Resting-State Networks in the Absence of the Corpus Callosum." *Journal of Neuroscience* 31 (42): 15154–15162.

Vogt, B. A. 2009. "Regions and Subregions of the Cingulate Cortex." In *Cingulate Neurobiology and Disease*, edited by B. A. Vogt, 3–30. Oxford University Press.

Vogt, C., and O. Vogt. 1926. "Die vergleichend-architektonische und dievergleichend-reizphysiologische Felderung der Großhirnrinde unter besonderer Berücksichtigung der menschlichen." *Die Naturwissenschaften* 14 (50): 1190–1194.

Volkow, N. D., J. S. Fowler, G. J. Wang, R. Baler, and F. Telang. 2009. "Imaging Dopamine's Role in Drug Abuse and Addiction." *Neuropharmacology* 56 (supp. 1): 3–8.

Von Bertalanffy, L. 1950. "An Outline of General System Theory." *British Journal for the Philosophy of Science* 1:134–165.

Wager, T. D., L. Y. Atlas, M. A. Lindquist, M. Roy, C. W. Woo, and E. Kross. 2013. "An fMRI-Based Neurologic Signature of Physical Pain." *New England Journal of Medicine* 368 (15): 1388–1397.

Wagner, G. P. 2014. *Homology, Genes and Evolutionary Innovation.* Princeton University Press.

Wang, G. H. 1965. "Johann Paul Karplus (1866–1936) and Alois Kreidl (1864–1928): Two Pioneers in the Study of Central Mechanisms of Vegetative Function." *Bulletin of the History of Medicine* 39 (6): 529–539.

Warrington, E. K. 1981. "Concrete Word Dyslexia." *British Journal of Psychology* 72 (2): 175–196.

Watts, D. J., and S. H. Strogatz. 1998. "Collective Dynamics of 'Small-World' Networks." *Nature* 393 (6684): 440–442.

Wikenheiser, A. M., and G. Schoenbaum. (2016). "Over the River, through the Woods: Cognitive Maps in the Hippocampus and Orbitofrontal Cortex." *Nature Reviews Neuroscience* 17 (8): 513.

Woodward, J. 2013. "II—Mechanistic Explanation: Its Scope and Limits." *Proceedings of the Aristotelian Society, Supplementary Volumes* 87:39–65.

Yasui, Y., C. D. Breder, C. B. Safer, and D. F. Cechetto. 1991. "Autonomic Responses and Efferent Pathways from the Insular Cortex in the Rat." *Journal of Comparative Neurology* 303 (3): 355–374.

Yeo, B.T.T., F. M. Krienen, J. Sepulcre, M. R. Sabuncu, D. Lashkari, M. Hollinshead, J. L. Roffman, et al. 2011. "The Organization of the Human Cerebral Cortex Estimated by Intrinsic Functional Connectivity." *Journal of Neurophysiology* 106 (3): 1125–1165.

Young, M. P., C. C. Hilgetag, and J. W. Scannell. 2000. "On Imputing Function to Structure from the Behavioural Effects of Brain Lesions." *Philosophical Transactions of the Royal Society of London. Series B: Biological Sciences* 355 (1393): 147–161.

Zangemeister, L., F. Grabenhorst, and W. Schultz. 2016. "Neural Basis for Economic Saving Strategies in Human Amygdala-Prefrontal Reward Circuits." *Current Biology* 26 (22): 3004–3013.

Index